PERSON AND GOD
IN A SPANISH VALLEY

STUDIES IN SOCIAL DISCONTINUITY

Under The Consulting Editorship of:

CHARLES TILLEY

University of Michigan

EDWARD SHORTER

University of Toronto

PERSON AND GOD
IN A
SPANISH VALLEY

WILLIAM A. CHRISTIAN, JR.

1972 SEMINAR PRESS New York and London

SEMINAR PRESS, INC.
111 Fifth Avenue, New York, New York 10003

United Kingdom Edition published by
SEMINAR PRESS LIMITED
24/28 Oval Road, London NW1

LIBRARY OF CONGRESS CATALOG CARD NUMBER: 72-7697

PRINTED IN THE UNITED STATES OF AMERICA

CONTENTS

PREFACE

This is a study of how communities and individuals relate to the divine in a mountainous valley of northern Spain. It is the first report in a series that will include the geography and life cycle of shrines in all of Spain as well as a study of apparitions in the Spanish Peninsula in the last one hundred years.

It is the product of the total time I spent in the Nansa Valley—twelve months over a two-year period, 1968–1969. It is augmented by my wider studies of shrines and apparitions, although these remain to be written.

The organization of this work is based on two principles. (1) By locating group identities and understanding the way they arise, one can begin to understand what role religious symbols play in the lives of different people. (2) By understanding the types of relationships existing among humans, one can begin to understand the relationships of those humans to holy figures.

The first chapter of this book examines the different group identities of the inhabitants of the Nansa Valley and searches for the activities and institutions that produced them. By studying the ecology and history of the valley in relationship to the rest of the world, one comes to understand the different classes to which each of its inhabitants belongs. The second chapter deals with the shrines and devotions that correspond to the different levels of identity from nation to family. The rise in popularity of some shrines and the decline of others coincide with shifts in identity of the people in the valley. The first and second chapters, which present, in turn, the people and their saints, become a backdrop for the third, which, among other things, evaluates the time spent in personal relationships as opposed to that with holy figures.

The underlying question is: What types of people and under what conditions are they most open to religious experiences? The first chapter is useful here for it describes in detail the characteristics of social relationships in the valley. The second chapter, in turn, provides the repertoire of holy figures available to the inhibitants of the valley. Based on this information and interviews with my friends, I have, in the third chapter, evaluated the quality of the individual's relationship to holy figures. Wherever possible, I have tried to add historical dimension to my observations on present-day society and religion in the valley. The study concludes with a rough schematization of the evolution, over time, of modes of conceptualizing human relationship to the divine.

The valley I chose to study was not chosen to be typical of all Spain. That would be an impossibility, given the wonderful variety of environments and the many different religious cultures. The rural parts of the province of Santander are among the most devout areas of Spain in terms of outward practice (which is the only measure of comparison available). In this valley attendance at Sunday mass is virtually complete. Almost every village has a son or daughter who is a priest or nun or religious somewhere. Virtually everyone in the valley over the age of seven receives communion at least once a year. And no one has died outside the faith.

The high level of religiosity is shared with the rest of the north of Spain along both slopes of the Cantabrian chain, the only exception being the mining regions of Asturias. It is a religiosity that probably accounts for the early literacy in the region, a literacy that in turn has augmented and deepened the religious culture. The earliest figures on literacy in Spain (the census of 1870) show Santander in a preeminent position. And a study of literacy in 1920 showed the Nansa Valley among the topmost 6% of all judicial districts in Spain.[1] The lady in whose house I stayed in San Sebastion read every afternoon from an edition of Fray Louis of Granada that had been in her family since the seventeenth century. Interviews with the citizens presented in the third chapter of this book demonstrate a very sophisticated level of theological information.

So to the degree to which religion is important, what I have learned will not be typical of all of Spain. But many of the configurations—the uses of the divine, relations with the divine, the attitudes toward divine images as symbols for group identities—are common not only to the rest of Spain, but also to much of Roman Catholicism as well. Some of these findings may be particularly useful as a point of departure for the study of Roman Catholic culture in Latin America when the difficult process of disentangling Spanish, Indian, and African origins is undertaken. So far as I know this is the first ethnographic study of Roman Catholicism in the European context.

The approach I have adopted owes much to the work of William James, Laurence Wylie, and Peter Berger. For this first part of my project my primary sources have been my friends, the people and priests of San Sebastian, Tudanca, Obeso, Cosio, and the other villages of the Nansa Valley. I owe special thanks to those who have lodged me and coached me in the ways of the culture, Candida Gonzalez, Agustin and Ana Grande Diaz, the family of Faustino Gonzalez Gomez, and the family of the late Manuel Agüera Vedoya. For secondary sources I am indebted to other Spanish scholars: those of the region—Jose María de Cossío and Tomas Maza Solano—and those of the nation—Juan and Rocio Linz, Carmelo Lisón Tolosano, and Juan Diez Nicolas.

I am grateful to a number of Americans who have helped me at some stage of this work, particularly the distinguished L. W. Bonbrake, who has been a constant solace, Robert Burns, William and Rena Christian, Magareth Chamberlain, Susan Harding, Max Hierich, James Lang, Howard Preston, Roy Rapoport,

[1] Census of 1870 and L. Lizuriaga, "Analfabetismo en España." Madrid, 1926.

James Robertson, Guy Swanson, Charles Tilly, Mischa Titiev, Eric Wolf, and Jay Wylie.

Work on this particular part of the study was partially financed by the Ford Foundation grant to the University of Michigan for Mediterranean Studies and by the Center for Research on Social Organization of the University of Michigan.

Noel Buckner, Piedad Isla, and Toñin of Celis have kindly provided photographs for the book.

William A. Christian, Jr.
Wilton, New Hampshire
and Tudanca, Santander

PROLOGUE

My interest in religion in Spain began in 1965, when I walked across the northern part of the country from the Pyrenees to Galicia. On that trip I could not help but notice the overwhelming devotion to Mary, in contrast to what I had known in my Protestant upbringing. Later, under Guy Swanson and Eric Wolf, I read what I could about the growth of devotion to Mary, and became fascinated with the legends of appearances and miraculous findings of statues in Spain. I subsequently visited most of the 300–400 regional shrines in Spain, learning about the kinds of devotion prevalent, gathering books and pamphlets about their history. Finally, to better understand the mode of worship, it seemed necessary to live in one place and study how Mary and the saints were regarded by a community and its members. In 1968 I chose the village of San Sebastian de Garabandal in the province of Santander because it was the site of apparitions of Mary and Saint Michael in the early 1960's, and I thought that the people would be more willing to talk about their relationships with the divine. I quickly found that in these mountainous regions every village is a world unto itself and religious devotion varies from one to the next. So I moved around, staying several months in San Sebastian and neighboring Tudanca. I also visited all the other villages in the Nansa Valley and studied the archives of all the churches.

At first I was seen as a visitor or as a writer. But as I participated more and more in the events of their everyday lives—playing the group games of the teenagers on Sunday afternoon at San Sebastian, haying and tending animals with the herdsmen, accompanying families on trips to shrines, sitting in the doorways and talking with the elderly, going to dances—I was seen more and more as an individual, and precisely what I was up to became less and less important. I myself felt more and more at home. Adopted by the villages that I lived in, I was treated with great friendliness and generosity, the kind of gentility that characterizes people who all, until the eighteenth century, were *hijos de algo.*

Because I was interested in the history of the valley, especially its religious history, and because I asked questions about shrines, people in the valley assumed that I was devout. I had taken some pains to dissociate myself from the pilgrims who were arriving from all over the Western world at San Sebastian because of the apparitions, and my dissociation was generally believed because I was a Protestant (and hence an unbeliever in apparitions). But because I talked to priests as well as to other people, and because of the subject matter of my inquiry, the villagers assumed that my interest in religious matters was more than academic. For this reason I believe that some of the men were hesitant to express disapproval of the Church to me, and on questions of personal attitudes toward religion I found I could only feel sure that I was getting true opinions and feelings from people I knew very well.

I owe my readers, as I owed the villagers, some statement of my beliefs in religious matters. I was brought up a cross between Congregationalist and Quaker. I would say that in times of trial or aloneness, like the villagers, I turn to my friends for help. And failing them, unlike the villagers, I turn inward rather than toward divine figures. I came to the study with few fixed beliefs but with an openness to the possibility of the validity of the beliefs of others. I believe I inherit this mixture of sympathy and objectivity from my father, who is a student of religion. My interest is in understanding what people do. Religion is one of the things that people do, for some of them the most important thing. Here I have tried, for a Catholic community in Spain, to understand the rhyme and reason of their relations to supernatural figures.

The steep green slopes of the Nansa Valley in the province of Santander, Spain, support an economy based on cattle. The villagers have most of their dealings with the divine through specially located images or shrines. These shrines correspond to levels on which the people form a community or have a sense of identity (nation, region, province, vale, village, barriada), and are used by women, men, and communities as an aid to the solution of specific problems. The nonshrine images in the parish churches and homes are of more recent vintage. They are used for individual and family devotions, and are especially turned to, particularly by the women, as aids to salvation.

The relationships within the family among father, mother, and children somewhat correspond to those among government, patron, and villagers and those among God, Mary, and believers. The mother, the patron, and Mary are essential mediating elements between the other two parties in their respective units. Partly because this analogy is so strong, persons in the valley turn to divine figures when they feel most alone. The divine figures such as Mary and the saints seem to stand in for missing human figures at times of intense personal need— during courtship, at widowhood, in sickness, upon emigration, and at death. On these occasions a person may fix on a particualr saint as a personal patron to whom he becomes accustomed to turn, subsequently, for comfort and assistance.

Parish death and baptismal records and images in the churches attest to a succession of popular devotions over the past 400 years. Their rise and fall have been caused in part by the proselytizing activities of the religious orders and the bishoprics and the regional reordering of economic life around major cities. The people and the priests have particularly concentrated their devotion on Mary in the past century, for she has served as a rallying point in the national and international struggles against secularization.

The human modes of exchange with divine figures (and ultimately, with God) parallel their modes of exchange with each other. The shrines are the major exchange centers where debts to the divine are paid. This practical kind of religion centering on shrines seems to be the oldest form in the valley—virtually the only form that most men participate in. It is integrated into the landscape, for the shrines have a supernatural rationale (generally through apparitions) for being located precisely where they are. They seem to be control points at which the people attempt to influence the penetration of foreign material and power into their countryside. This religion divides time and space into the sacred and the profane. Its ideological charters are the origin legends of the shrines, which confirm special divine favor to a given community through a given image at a given location and time of year.

A later kind of religion, which may have entered only after the Council of Trent, has served to inculcate a sense of sin and fear of purgatory in the villagers; it encourages them to adopt salvation as their ultimate goal. This kind of religion, which divides human states into the pure and the impure, has had its greatest impact on the women. Its root metaphors are the story of the expulsion from the garden of Eden, on the one hand, and the resurrection of Christ, on the other.

The religion being brought into the valley by the young priests of the Second Vatican Council attempts to encourage people to find God in each other and do away with religion for practical purposes or salvation. Their approach erases the need for divine intermediaries and questions the continued use of shrines and generalized devotions.

William A. Christian, Jr.

CHAPTER ONE

THE PEOPLE: ACTIVITY AND IDENTITY

Adiós Reinado de España
Adiós Valle de Rionansa
Provincia de Santander
Nobleza de la Montaña.

Adiós Pueblo Rozadío
Donde yo pasé mi infancia
Adiós mi padre y mi madre
a mi cuñado y mis hermanos.

Goodbye Kingdom of Spain
Goodbye Vale of Rionansa
Province of Santander
Nobility of the Montaña.

Goodbye village of Rozadío
where I spent my childhood
Goodbye father and mother
brother-in-law, and brothers.

from the epic trova of
Manuel Agüera Vedoya, 1926

I. Introduction: The Annual Cycle

Entre estas cosas fue creciendo mi ánimo. Los hitos del tiempo eran los motivos trascendentales de la Naturaleza en lo anodino del pueblo. Épocas de nieve, de cosecha, de trajín en las tierras, de ocio en los portales y en las cocinas . . . . No se decía la primavera, el verano, el otoño, el invierno. Se decía la época de los vendavales, de la caída de la hoja, de las golondrinas, de las cerezas, de la siega, de las panojas, de las nueces, de las magostas, del ábrego. Todo el tiempo sin los hitos numéricos del calendario. Cronología marcada por los aperos, por la nieve, por el viento, por las romerías, por las novenas, por las costumbres, por los pájaros trashumantes.

Among these things my spirit was growing. In the dullness of village life the milestones in time were the momentous changes in Nature. The epochs of snow, of harvest, of work in the fields, of idleness in the doorways, and in the kitchens . . . . They did not speak of spring, summer, autumn, winter. They would speak of the time of the gales, of the falling of leaves, of the swallows, of the cherries, of the haying, of the maize, of the walnuts, of the chestnut roasts, of the south wind. All the time without the numerical markers of the calendar. Time marked by the tools, by the snow, by the wind, by the pilgrimages, by the novenas, by the customs, by the migratory birds.

from *La Braña,* by Manuel
Llano, (1934)[2]

1

The Cantabrian mountains stop the rain before it reaches most of Spain, leaving most of the country arid and brown, but the narrow 50-mile strip of countryside between the mountaintops and the Atlantic is lush and green. Stretching from the Basque country in the East through Santander and Asturias to open out into Galicia in the northwest corner, this strip is part of a zone of high rainfall that goes up the European coast through Norway and includes much of Ireland, Cornwall, Wales, and Scotland. It is known to cultural ecologists as The Atlantic Fringe. The weather limits the agriculture fairly strictly to the raising of livestock, for cereals are impractical with so much rain.[3]

This relatively homogeneous horizontal strip along the Bay of Biscay is divided vertically into a number of grooves formed by rivers draining the rainfall into the sea. There are about thirty such valley systems from the Basque country through Asturias. Since roads normally go up the valleys and towns are located on the sides or floors of valleys (depending on whether their chief activity is herding or trading, respectively), most valleys are distinct political and cultural units. At various points the valleys narrow and these points serve to mark off the valleys into townships or vales. (See Fig. 1.)

The Nansa is one of three rivers (the others being the Saja and the Deva) that run to the Bay of Biscay in the western portion of Santander. I became most acquainted with the way of life of the three upper vales of the Nansa: Polaciones, Tudanca, and Rionansa. These three vales had nine, four, and six villages, respectively. With minor variations due to differences in rainfall, soil quality, and slope, the way of life of these villages applies to most of the upper portions of the provinces of Santander and Asturias.

Each village is in a cup of hills. The valleys are steep and green; one day out of four brings rain, and often shafts of sunlight come down through broken clouds and illuminate a field, then move across the valley and over the mountain. Rainbows are common, and the many colors of green in the valley are met by many different colors in the sky. The rain gives life to everything. The hillsides begin the spring brown, then green, then brighten with the yellow of the gorse; all is green all summer. In the fall the ferns turn first russet, then golden, then golden on white as the first snows come.

As if to parallel the changes in the scenery by the seasons, there are several fairly consistent moods to be found among the people at different epochs in different settings. That of the upper pastures in spring and fall, free from the village, is airy, open, and honest. Among the herdsmen there is an easy freemasonry that betokens an escape from the village. Food is shared and eaten from common bowls; tools and stable paraphernalia are freely borrowed. Some of the men sing as they work, breaking off into long cadenzas as they drive their cows before them or climb up the mountain into a cloud.

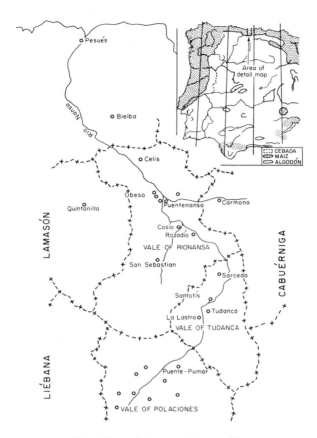

Fig. 1. Map of the upper Nansa valley.

The opposite of this is the mood of the village center in the fall, winter, and spring, when the teenagers are away on their seasonal jobs. For life is more difficult at close quarters. There are people with whom, for one reason or another, one does not speak. There is an undertone of competition. Conversations are likely to be about other people. The social circle is more restricted in space, to the hearth or the cafe. The village at times acts in the same way as it is arranged, a lot of people facing away from each other, huddles of houses back to back.

In the summer when the teenagers return from their jobs in the lowlands to work the hay, there is a bustle that partakes both of the freedom of the fields and the tension of the village. The houses awaken; families eat picnic lunches in the pastures; teenagers visit from family to family; and people call greetings and sallies to each other as they pass. The work partially diverts people's attention from each other.

Finally, there is the fiesta. For the village holiday, lasting two or three days every year, there is a respite from work and a truce from disputes. Relatives return, people visit from other villages, and for a time the village opens up. The fiesta is run by the teenagers, who are removed from the problems of their parents; problems stemming from disputed inheritances and boundaries, politics, and broken engagements. Most of the youth are away for half the year, free to work elsewhere, to meet new people. Their parents must stay in the village, must work out ways to coexist with their neighbors in a small, concentrated community.

Each vale is a township, and the townships are divided into *concejos*, or councils, which are the governing bodies of the individual villages. In the Nansa valley every village's land is used in four ways: the cornfields, gardens, meadows *(prados)*, and wilderness *(monte)*. On the valley bottom and sheltered flat places, maize is grown. These lands, usually close to the villages, themselves in sheltered positions, are known as the *mies*. They provide food for the villagers and some of their animals. The maize is ground into corn meal from which flat corn cakes *(tortas)* are made. Before maize was brought in from America (it was cultivated in the valley at least as early as 1626) rye may have been grown. Until the mid-1950's homemade torta was the staple food. Now corncakes are made only by herdsmen tending cattle in the mountains, for virtually all the familes can afford to buy bread. The maize kernels are fed to the fowl that each family keeps, and the stalks are fed green to the cattle in the fall. Because maize is coming to have less utility, many families in the valley villages are leaving their mies fields fallow and cutting them two or three times a year for hay, or cultivating artificial pasturage like alfalfa. Other vegetables—generally potatoes and occasionally beans—are grown in the mies, but not extensively. Families also grow limited amounts of fruit and vegetables which require closer attention in gardens close to their houses. These gardens are often walled to keep out animals and children.

Uphill from the mies are private meadows that produce hay for the winter. The hay for the meadows close to the village is kept in lofts in the village, and hay from the meadows farther than a five-minute walk from the village is stored in winter barns called *invernales*. The fields are often divided among many owners with stones marking off the different strips. The long stone barns also may belong to as many as four or five owners, each with a section of loft and stable.

Finally, around the hay fields is the land unfit for cultivation referred to as *monte*. Some of it, gorse- and fern-covered mountainside, is fit for sheep and goats, and communal or private flocks graze it. The more remote corners of the village territory may still be forested, owned collectively by the village and occasionally auctioned off for logging when the commune is in financial need. The forest, seemingly unproductive, is an essential part of the countryside. The woods provide firewood for the houses, and timber for hay sleds and wooden implements. Previously they also supplied lumber for the frames of the stone houses, although now much ready-cut lumber is brought in from outside the villages.

The highest land, along the ridges and on the mountains, is excellent pasture land. Here the cows and horses are sent during the summer while the meadows are being harvested.

Throughout the alpine and lower mountain zones of western Europe there has been a slippage of meadows reverting to wilderness—*prado* to *monte*—in the last twenty-thirty years.[4] This has occurred for two reasons: First of all, meadows left uncut for several years quickly revert to gorse and fern. As more families move out of the mountain villages there are not enough arms to cut the hay, and the least desirable fields lapse into the savage state. In the Nansa valley there are many such fields half-eaten by encroaching scrub. The second factor arises from absentee landowners: villagers who have moved to the city, the Ministry of Reforestation, or even occasionally nonherdsmen still living in the villages who plant tree seedlings on plots of field. The young forest quickly encroaches on the fields. The two cannot exist side by side because the roots from the trees spread under the grass and take up the moisture. Behind both these phenomena of reversion and reforestation lies the fact that cutting the hay by hand is uneconomic. For the effort involved the return is pitiful, and there are fewer and fewer men willing to spend time cutting marginal fields. I should say fewer and fewer sons. The fathers of local families are in the herding enterprise for the duration of their active lives, but fewer and fewer of their sons, once they pass a certain age, are willing to come back in the summer for the haying or to take over the fields and barns when their fathers retire.

In the Nansa valley there are three kinds of cows: the local Tudanca breed, a tough animal capable of climbing and surviving on the steep mountains, which is sold for meat; and two varieties of milk cow, Swiss and Holstein. A *cabaña* (the sum total of a herdsman's animals) might include both milk and meat cattle, the proportions varying according to altitude in the valley. The higher villages, those of Polaciones, Tudanca, and Rionansa, have an overwhelming predominance of Tudanca cows, with occasional milk cows for family use. Until recently this was the pattern for virtually the entire province. The villages nearer the coast would pay to the higher villages pasture rights for upper mountain pasturage, while they harvested their own meadow hay. But with the advent of the train and the automobile and the growth of an urban market for milk since the Civil War, the lower villages have converted their herds almost exclusively to milk cows, and trucks come to pick up the milk every morning. As a result they have fewer cows, but the cows are worth more. With a reduced number of cattle and intensive cultivation of artificial pasture, they now can make do with the lands around the village, so few farmers from the lower Nansa valley have herds of cattle that they send up to the mountains for the summer. The changeover from beef cattle to milk cows has had a profound effect on these villages, accelerating the disintegration of the oldtime communal solidarity.

The upper villages cannot convert to milk cows for several reasons. One is that the villages are that much farther from the coastal cities, the markets for the milk. But more important, the meadows of these villages are too far from the road, and the paths from the meadows too precipitous and rough to permit the easy transportation of milk from the barns to the village or the daily driving of cows from the fields to the village for milking.

As in most herding cultures, the people of the upper Nansa valley have a certain amount of moving to do. Early in the spring, when there is enough grass on the mountainsides and it is warm enough to leave the cattle out all night, the cows are loosed on the common grazing ground of the mountains. Each herds- man has a sector where he puts the cows, so he will know the general vicinity when he goes to check on them. Checking on the cows throughout their stay in the mountains is necessary for two reasons: If a cow or calf falls from a cliff or breaks a leg in a hole and the herdsman discovers it in time, then the meat can be saved; secondly, the herdsman must keep his cattle from straying into the land of neighboring villages, where they can be captured and held for fines.

Cows that have borne calves, calves soon to be sold, and the milk cows are kept back in the winter barns where there is still hay. In these pleasant days of early spring the herdsmen arrive early in the morning from the village, turn the remaining cows out to pasture, push the manure out of the barn, and spend the day talking with each other, making wooden shoes or tools, and checking on the cows in the hills nearby. At dusk the herdsmen bring the remaining cows back into the barn, feed them, milk them, and go back to the village. These are the easy days.

In early May the schedule is reversed, since the man is needed to plough the mies. He may spend the night at the winter barn and the days in the village. At the same time other members of the family transfer the manure piled outside the winter barns to a boxlike contraption on the family burro, and thence to the nearby meadow where it is spread as fertilizer for the new grass.

By the beginning of June all remaining cows except the milk cows are turned loose to graze in the backlands. Those farmers with large herds and enough capital find it profitable to take their cattle up to higher valleys where the grass is more nourishing. The entrance date for these valleys is generally San Antonio (patron of animals), June 13. On this date long files of cattle converge from many lower valleys on the high mountain valleys of the province of Palencia above Lores and the district of Campóo in the province of Santander. There the proprietors of the valleys mark down the number of cattle being entrusted to their care. A flat fee per head is paid, and in the more isolated valleys a shepherd is hired to keep track. I accompanied a herd of 40 cattle from

Cosío on a three day drive to a summer pasture in Palencia, a total of 50 kilometers. The advantages of sending one's cattle up to rented pasture are that the grass is better and hopefully the rental fee (varying from 200 to 550 pesetas per head in 1969) will be regained by having a heavier, healthier animal in the fall. Also the local common lands tend to be somewhat overgrazed now, as there are more cattle than ever before. Another advantage is that the herdsman does not have to make periodic checks all summer when the cattle are being tended in the upper valleys; rather he can devote all his time and energy to haying.

Once the cows are out of the way, it is time to cut the hay in the meadows *(Hacer el verano)*. This is the major task of the year and in the upper villages lasts from June until September, October, or November, depending upon the delays caused by rain and overcast weather. It is the only reason that sons and daughters must come home every year. The amount of hay a family can take in during the summer limits the number of cattle they can keep over the winter. Hay is everything. Surplus hay can be sold for a good price. A lack of hay might mean that the herdsman would have to liquidate some cattle at market under unfavorable conditions or that all cattle might suffer from undernourishment. Consequently these summer months are times of intense activity and a certain amount of anxiety in the villages. The work is hard, and the fear that there will not be enough sunshine means that every moment of sunny days must be used to the fullest.

The cutting of hay is done by the men and older boys; perhaps scything is the most skilled work involved. In addition, sharpening the scythe by hammering on the edge and then using an oilstone have been raised to art forms in the villages, and different villager's styles are analysed and appreciated. In the wake of the harvesters follow the women and children, turning the grass with rakes and making piles when it is dry. Then the grass is carried to the barns. Customs of carrying hay vary from village to village, depending mainly on terrain. In the lower villages the hay is carried to the barn by oxcart; in the villages with steep slopes it is carried in heavy bundles (called *coloños*) on a man's back, on horse sleds, or on ox sleds (called *basnas*).

Each family owns and uses fields and barns in several parts of the countryside. The mowers start with the lowest land, which, being free of snow the earliest, has the tallest grass first. This land is likely to be in or around the village—in the mies or nearby. Then they move progressively to higher ground until they complete the highest meadows. Finally they move back and cut the lowest fields once again. Manuel Llano, the writer from Carmona who opened this introduction, describes the mood of this summer work in his novel, *El Sol de los Muertos*:

> El alimento de las bestias en los días crudos de la invernada movió todas las actividades, fatigó todos los brazos, sofocó todos los rostros.

Aquel incesante rumor de dalles y de rastrillos, de picos y de piedras, aquel ir y venir de las accareadoras con el enorme coloño a cuestas; aquel encorvarse de las que escogían los helechos de las lombilladas; aquel cantar quejumbroso de los carros; todas las características de los trajines estivales en las extensas praderas de Llendejosó, fueron a manera de un intensísimo apogeo del trabajo campesino, bajo el fuego del sol.

Antes del amanecer, ya está el segador encorvado sobre el yerbío, desnudos los brazos y el pecho, quemada la faz, mojados los cabellos y doloridos los músculos. Cuando el sol se ha ocultado tras las crestas, perpetuamente nevadas, aun se oye el rumor de los dalles en la pradera, entre las canciones de las mozas que vuelven al lugar, con el apero al hombro.[5]

The feeding of the animals in the rigorous days of winter moved all the activity, tired all the arms, flushed all the faces.

That unceasing sound of scythes and rakes, of hammers and sharpening stones; that coming and going of the carriers with their huge load of hay on their backs; that bending down of those taking the thistles out of the cut hay; that plaintive song of the carts; all the characteristics of the summer labor in the extensive fields of Llendejosó, served as a most intense climax of the work of the countryside, beneath the fire of the sun.

Before dawn already the mower is bent over the grass, his arms and chest bared, his face burned, his hair soaked and his muscles aching. When the sun has hidden itself behind the ridges, always white with snow, still can be heard the sound of scythes in the meadow, beneath the songs of the girls who return to the village with rakes on their shoulders.

Until recently, when the fields were cut for the last time in the late summer or fall, they were opened to all village animals: sheep, cattle, and horses. The cattle at the end of the summer were brought down from the upper valleys and back mountains and allowed to browse over the cut fields and the mies. This was known as the *derrota*. It conditioned the entire summer schedule, as it required a certain coordination of labors, and it now occurs only in very few villages. In San Sebastian and Obeso it was abandoned in the mid-1960's with the general waning of communal activities.

The day after the last hay is stored in the barns, all of a sudden, the teenagers leave the village for their winter jobs, the children are in school, and the village is quiet for another year. About a month later, in October or November, the village turns again to the mies; for a few days the calm is interrupted, and some of the teenagers return to help cut the corn stalks and carry them back to the village houses in bundles. There the ears are strung up to dry on porches or in the kitchens, and the stalks are fed to the cattle.

Each morning during the winter the herdsman goes to the winter barn where the cows are, turns the cows out to water and graze if the ground is not covered with snow, and then supplements their diet with hay from the loft at eventide. The order of use of the winter barns is intended to minimize the amount of circulating which the cows are called upon to do. Thus they are often taken first to the barn closest to the summer pasture, then closer to the village, and then back up toward the summer pasture as Spring approaches. Extreme care is exercised in calculating the amount of hay to be handed out so enough will be left for the last critical months of March and April.

In sum, the people of the village rotate around the village in the back country in the summer cutting hay, the cows rotate around the village eating the hay in the winter.

Because the landscape is suitable for any agile grazing animals, some cabañas include a small number of sheep, enough to provide the family with lamb on festive occasions, wool to sell, and milk for cheese. Again, until recently sheep were tended in communal flocks, either by a paid shepherd or by rotation among the villagers. At present this system holds only in a couple of villages in the vales of Tudanca and Polaciones. In the other villages, only those with many sheep (say 10-15 as a minimum) keep sheep at all. An elderly couple with few expenses can live with an income from a herd of 75-100 sheep, and for a household with only semiactive workers (children or oldsters) shepherding is an alternative to cattle herding. One need own no land, as the village backlands are communal property.

Another alternative is goats. At times there have been more sizable goat herds, but at present there are only two or three small herds in the valley, providing supplementary income for cattle herders.

A third alternative or supplement to the basic industry of cattle raising for the townships with much free grazing land is horse raising—the horses being sold for meat. Few men raise horses as a major source of income, but horses have advantages over cattle in that they can remain out all winter except in the most rigorous conditions and they need little tending.

In addition, many families keep chickens for eggs and a pig that they slaughter in the fall or winter for sausages and hams. Many herdsmen keep a dog, and in villages in which ox carts or sleds are used to transport hay, each active household has either a pair of oxen or a pair of cows that have been domesticated for work as a team.

A number of persons in these villages hold occupations other than herding. About 20 miners, generally exherdsmen or the sons of herdsmen, live in the lower villages of Rionansa and work in the mines of La Florida, just outside Celis. Their children, not having to work in the daily round of tending animals, will not be or marry herdsmen, and thus probably will not stay in the valley.

The shopkeepers usually come from farming families and often maintain a few animals on the side. Except in Puentenansa and Cosío, which have some specialty shops run by widows, all the stores in the valley are part store, part tavern. On the average there is one such store for every 20 households in the valley, but they are very much more frequent on the main road than elsewhere. The storekeepers' children, like those of the miners, may receive a better than average education. The sons do not go back to herding. They either stay on in the valley as shopkeepers or move on to work in the cities. Very rarely will a daughter of a small shopkeeper marry a herdsman. The shopkeepers are closer to the villages and their problems than the professionals, although the villagers resent them for taking their money.

The valley also has a thin stratum of professionals. Every couple of villages has a priest, and every vale has a doctor and a municipal secretary who expedites most official matters. Every two vales has a veterinarian, and a small number of persons work for the electric company as engineers, tending the power stations at La Lastra and Rozadío. Finally in Puentenansa there is the upper valley's only pharmacy, and a Civil Guards barracks, whose officers participate in the society of professionals.

The professionals associate with each other fairly exclusively. They have no animals, tend no land. Only exceptionally is one of them from the valley. All have automobiles and make frequent trips outside the valley. Their children are educated in schools away from home. Because they are not born in the villages, they cannot attend the village government meetings.

Their number is augmented in the summer by an occasional summer family. Virtually every village has its quota of successful sons who come back for short stays in the summer from Madrid, Santander, or the New World.

Finally there is an occasional nobleman. Each village has a couple of large houses or manors. In the smaller villages as often as not they are lived in by farmers, having been lost by indigent nobility, or the once noble family having fallen to the common level over time. In two or three villages the manors (called *casonas*) are inhabited part of the year by wealthy hidalgo families. Until the late eighteenth century, virtually all the inhabitants of this valley were hidalgos, like the inhabitants of the Basque country. But at that time the ranks of the hidalgo class were purged, and only those with enough wealth to enable them to live without manual labor were allowed to keep the noble title. In practice this often meant those who had gone to the New World and come back wealthy. In the nineteenth century such families generally held large tracts of land and maintained herds of cattle. Some of the noblemen, the more charismatic, as in Pereda's novel about Tudanca, *Peñas Arriba*, had a great effect on the political climate and exerted much power in the villages. Now the land is sold and what nobility there is is almost in the same category as summer people.

All in all, the nonagricultural proportion of the population in these villages varies between 5 and 15%. The towns with the highest nonagricultural proportion are Rozadío and Puentenansa, the former because of the electric power station and the latter because it is a commercial and administrative center. The villages in the valley vary in size from 10 to 60 households. The population is extremely difficult to measure because there is so much seasonal variance. The population per household may average as much as five in the summer, and as little as 2½-3 in the winter. In other words the population of the smallest village varies from 25 in the winter to 50 in the summer, and the population of the largest villages varies from 150 in the winter to 300 in the summer.

Our remarks in the rest of the study apply generally to the agricultural population, unless otherwise noted.

II. Identities: National Identity

Human behavior and human communication does not form a uniform net across a landscape. It is more often clustered and discontinuous. In the process of introducing the pattern of life and social relationships in the Upper Nansa Valley, we will try to isolate the relevant cognitive boundaries of the inhabitants. We will look, on the one hand, for geographical areas in which the actual social communications are relatively more dense and, on the other, for the cultural identities which signify a sense of belonging or community among a group of persons.[6]

When searching for the "breaks" in a geographical system, certain clues are always helpful: the way persons group other persons and give them a name; the different sides that line up in conflicts; the uses of boundaries to isolate, or the relative ignorance of boundaries; the activities that weld groups together; and the ceremonies and symbols that persons share. Shared identities must be based on shared activities or a shared code for interpreting activities. Either the code itself or the actual activity provide a common experience. People can either be related because of proximity, and hence interaction, or because they share the same political or cultural hub, the same capital, which for economic or political reasons, in short because of the power and control it exercises over the lives of those in its hinterland, demands that all activities on the periphery be translatable into the terms of the center (e.g., all Spaniards must understand Castilian, use the decimal system, and participate in the capitalist economy).

People have an identity *as* something, as a Spaniard, for instance. They also have this identity *with* something, with other Spaniards. That is, people often share identities with others. Whatever it is that is the source for the identity links members of an identity group together. It provides a self-concept and the potential for a relationship with others who share that self-concept.

The study of what brings people together and what marks them off from each other is essential for understanding their relations with God, both collective and individual. Such a study provides an inventory of the groups that might want a divine figure to treat with or be protected by in their joint endeavors. Also, the uncovering of bases for the cohesion of groups helps to explain the nature of any given group's relations with the divine. And for both individuals and groups, a study of the matrix of human relations provides the context into which relations with the divine must fit.

Different religious figures have come to stand for the different identities that people of the valley share. The linking of a religious figure to a shared identity seems to have this effect: It elevates or generalizes the basis for identity to the status of a family relation under the love, authority, and protection of a divine parent.

The particular shared experience or code that is the basis for an identity determines the extent to which the community is bounded and delimits the boundaries within which its religious symbol will operate. The quality or intensity of the shared identity of any group will determine the degree to which the shrine and the religion are successful in inducing a sense of brotherhood in the group. For these reasons the study of identities and the activities that are their basis is a fundamental introduction to the study of religious attitudes.

This study will sketch the gradual focussing of the regional identity on the city of Santander; the minor effect of anything like a subregional cultural unit, the equivalent of a *pays*; the relative unimportance of the vale; and the overwhelming importance of the two most bounded and most concentrated spheres of activity and identity in the lives of these people: the village and the family.

Perhaps because the identity of the Montaña as a region is relatively weak, the national identity of the people in the valley seems to be somewhat stronger. In school and church this identity is continually reaffirmed. For instance at what might be termed the symbolic climax of the village year, the elevation of the host during the high mass said at the fiesta, a small band in the church balcony plays the national anthem. Less symbolic, but doubtless more effective is the daily influence of radio and television, which is the programmed essence of the national community.

As an identity based on common activity the national identity is clearest in the military service each man must perform for one or two years of his life. Many of the men in the village fought for Spain at one time or other—in the Moroccan Wars or in the Civil War. Similarly in an American village, although on a less organized basis, the veterans of war are carriers of allegiance and loyalty.

At a more fundamental level the national identity is based upon the language: the villagers can go to Torrelavega, Santander, Madrid, or Barcelona and speak to and be understood by others of the same nation.

For over a thousand years the people have spoken Castilian, and most of these elements of national identity are not new. The schools and the church, both of which emphasize allegiance and nationhood, are institutions of antiquity. And there have been other wars before those in this century. To begin with, men from this valley took part in the reconquest, as a sixteenth century stone marker on a house in San Sebastian notes. The vast majority of citizens listed in the reports of the Marquis of Enseñada in the 18th century are listed as *hidalgos*, exempt from certain forms of taxation because of service to the king.[7]

Yet there have been changes that have reinforced the national identity. Increased geographical mobility in the past decades; more frequent contact with foreigners; and above all the introduction of radio and television have brought interest in national and international affairs to a day to day level.

III. Regional Identity

This valley of herdsmen lies on the western periphery of the Montaña region of Spain, Old Castile's window on the sea. The language is Castilian, with minor variations; yet what the inhabitants call Castilla is the land over the mountains to the south. They are of something else. One clue to identities lies in diocesan boundaries, for the provinces of Spain date only from 1833. The first census, in 1769, was arranged by diocese. Although it is now entirely within the Diocese of Santander, the valley once lay in three different bishoprics and bordered on a fourth.

The highest section, the vale of Polaciones, was part of the Bishopric of Palencia until 1949. What little exogamy there was allied it more with the communities over the mountains in the province of Palencia than with the other villages in the Nansa valley. In fact the road that links Polaciones with the rest of the valley is quite recent. In the 18th century the market centers for Polaciones were Potes, to the west, and Cervera de Pisuerga and Saldaña, in Palencia to the south. Its external commerce also faced south. In the long winter months the villagers manufactured carts and cart wheels to sell in Castile, and in the summer large flocks of merino sheep were brought to its pastures from Extremadura. The only commerce with the villages to the north was the maintenance of a few cattle from lowland villages like Lamadrid. Even now there are remarkably few intermarriages between the "Purriegos" and those lower down the valley.

The middle section of the Nansa valley was under the jurisdiction of the Marquis of Aguilar in 1753 when a survey of its villages was made. Tudanca, Sarceda, Santotís, and La Lastra paid tithes both to the Archbishopric of Burgos and to the Benedictine monastery of San Pedro de Cardeña in Burgos. The next villages—San Sebastian, Cosío, Cabrojo, Celis, and probably Obeso—paid tithes to Burgos and to the Collegiata of Santillana, near Santander to the east. In the earliest times Santillana was probably the urban hub for this part of the valley, with some attention paid also to the port of San Vicente de la Barquera.

Lower down the valley, the towns of the vale of Herrerías were under royal patronage and paid tithes to Burgos, while the hamlets at the very mouth of the Nansa, which were in the domain of the Marquis of Aguilar, paid their tithes either to the Count of the Vega de Sella (those to the west) or to the church of San Vicente de la Barquera, which sent priests out to care for the hamlets around it.

Consequently in 1753 when the survey of the Marquis of Enseñada was made, the top of the valley faced Castile and Palencia, and the rest of the valley was the far limit of the Archbishopric of Burgos. At this time, the valley of the Deva river, then known as the Province of Liébana, which bounds the Nansa valley to the west, pertained to the Bishopric of León.

Things became even more complex when in 1757 the Diocese of Santander was created. At this point all of the valley previously in the Archbishopric of Burgos was switched to the Bishopric of Santander, with the exception of certain lower towns, like Bielba, which were turned over to still a fourth diocese, that of Oviedo.

Hence the valley formed by the Nansa river is not, historically, a single cultural unit, unlike Liébana, to the west. There is, indeed, no name for an inhabitant of the valley, per se. In 1753 the agriculture varied considerably with altitude down the valley. In Polaciones bread was made from home grown rye. In Rionansa and Tudanca, rye was not grown; all bread came from maize. Lower still, the amount of cattle herding declined sharply, until in the hamlets of Val de San Vicente, few families even had one cow, but instead farmed the rich lands with vines, fruit trees, and maize. Even today the valley looks in three directions. Many of the mountaineers still feel an affection for Asturias and Asturianos, although they do not identify themselves as such. Those of Polaciones are still more attached to Palencia, but the majority does indeed look to Santander and Torrelavega as their cultural hubs.

This change came about gradually over the past 200 years. The beginnings can be seen in the reconquest of Spain, which moved out from centers in the north, supposedly with the aid of soldier-monks. Already, then, there was a community of interest with the coastal region. But the coastal region to the north had no focus, and it was only as the cities of Santander and Santillana, both built around important monasteries, developed that these mountain valleys could be weaned away from Castile.

The city of Santander (San Emeterio) did not begin to thrive as a merchant seaport until the 14th century, and it was not made into an episcopal see until 1757. Only in 1827 when the provinces were created was Santander made the capital of an administrative region. Before then, most of what is now Santander had been a subdistrict of Burgos, with the coastal town of Laredo as its capital. Finally, at the turn of the century Torrelavega began its rapid industrialization, and the Nansa valley clearly came within the orbit of these two towns to the northeast. It was at this time (1899-1902) that a road was built up the valley to Tudanca, superseding an old stage route, and a bus line was established that linked the valley with the new railroad at Pesués, by the sea. School teachers, priests, doctors, and veterinarians came to the valley from Santander. And with the simultaneous construction in 1945 of the dam between Polaciones and Tudanca and the road through the narrow chasm, even Polaciones was brought into the Santander hinterland.

In practical terms, this development meant that many persons from this valley would go off to seek their fortune in Santander or Torrelavega during the winter months, instead of Castile or Andalusia. The townships were occasionally referred to in Santander's *Diario Montañes* and *Alerta*. The people in the valley could legitimately refer to themselves as Montañeses. But in spite of this demographic and administrative affiliation with Santander, there is little sense of belonging to the province as a community of interest; there is little affective identification with *La Montaña* in the valley. Not a tenth part as much pride as is exhibited by an Asturian to the west; nor a hundredth part as much pride as is manifest in a Basque to the east.

For in the romantic cult of regional identity that swept Europe in the nineteenth century in a reaction to industrialization and as a result of its newly developed regional cities, Santander had a tremendous handicap compared to its neighbors. It had no language, unlike the Basques; no claim to the heartland of the reconquest, unlike the Asturians. Aside from being a pocket of resistance to the Romans, it has no claim to historical separateness from Castile. Hence although the valleys and villages of Santander had their own individualities, some of them celebrated in the novels of José María de Pereda, Santander as a region had little or none. Even though the concept of Santander or La Montaña dates back at least 300 years, circumstances do not seem to have favored the forging of an emotional bond. The bond is far more restricted: To the vale and to the village, to the cup in the mountains, bounded by mountains, that is home.

There is a regional identity of a more diffuse nature that is not based upon a fixed region. It is the diffuse sense of a community of herdsmen arising from a shared occupation and the way of life deriving from that occupation. Such a regional identity is perforce limited to the herding region and hence includes neither the fishermen of the coast nor the Castilian dry-farmers. Its meeting places are the fairs from Reinosa to Potes. It would include the regions of Campóo and the mountains of Palencia to which some of them take their cattle, as well as the lower villages toward the coast who bring their cattle up for the summer.

IV. The Vale

The simplest way to establish the meaningful units of identity for the people in the valley is to search out the way they lump their neighbors together and then to see what is left over. In the course of my stay in the villages of San Sebastian, Tudanca, Obeso, and Puentenansa and in markets, fairs, and dances up and down the valley, I listened for references to geographically defined groups of persons, for the categories that made up the local cultural geography.

One step above the most important level, that of the village itself, was the level of the vale. By vales I mean the sections of the river valley that are more or less geographically defined by constrictions of the valley walls.

Each village considered the vales and villages higher in the mountains (with the exception of Polaciones) to be more rustic—and with perfect reason. The nadirs of rusticity were considered to be San Sebastian and Tudanca. Both those below them and the Purriegos above them considered them to be more uncouth than themselves. Both considered each other to be more *bruto*, themselves to be more *noble*. Polaciones, looking to the south, saw itself as closer to the centers of culture like Cervera, Palencia, or Madrid; the villages of the lower Nansa valley, looking to the north, saw themselves as closer to the centers of culture of Torrelavega and Santander. A very practical measure of the reality of these attitudes can be seen in the rate of exogamy, which is lowest in these villages at the end of the line.

The vale as a unit is more a functional reality, however, than just a simple categorization. First as fief, then as township, it has been an administrative district for centuries and as such has developed ways of sharing land and services that help to tie its villages together. Polaciones has perhaps the strongest identity as a vale. Many of its villages are so small as to make endogamy virtually impossible, and thus nearby villages are tied together by a net of kinship. In addition, all the towns in Polaciones share its mountain pastures and in the past shared the benefits of the rentals. When merino sheep were brought to pasture there the rent was paid (5000 reales in 1753) to a junta for the entire vale, which used the money to pay for goods and services needed for the vale as a whole. All up and down the Nansa valley, in 1753 and now, professionals are paid for by groups of villages. In 1753 the professionals so paid were doctors and notaries. Now they include doctors, veterinarians, and township clerks.

The main local meeting grounds for vales are the local fairs and markets. The main market in the Nansa valley is that of Puentenansa—a cattle market on the last Friday of every month from October to May. Cattle are brought to this market from the entire Nansa valley and the neighboring valley of Lamasón to be looked over by buyers from Torrelavega, Reinosa, Campóo de Suso, Cervera, and occasionally, the city of Palencia. The notables of the villages—the priests, town secretaries, police officers, veterinarians, and others with time on their hands—show up on this day to chat or play cards in the bars.

There are fewer fairs now than in the past. Each vale formerly had at least one fair toward the end of summer. Sometime toward the end of the nineteenth century the Tudanca fair was discontinued, and sometime in the twentieth century the fair at La Lastra that succeeded the Tudanca also disappeared. At present there are fairs in the upper valley in Polaciones on two dates in September and in Rionansa on Saint Michael's day (September 29). The big fair on San Miguel draws cows not only from Rionansa, but also from Lamasón, Carmona, and Tudanca. To the east Liébana has its own fair on All Saints Day in Potes, and some cows and horses of the Nansa valley are taken there. Similarly, cows from Tudanca are taken to the fair in Correpoco above Cabuérniga in July and to Terán in Cabuérniga in August. The fair provides a concrete occasion, eagerly looked forward to, for families in the vale to get together. The long day is passed buying drinks for old friends, comparing notes on the summer's progress, a picnic lunch, games of chance, a singing competition, and finally a dance.

Yet the zone delineated by the market and the fairs where cattle are sold is larger than merely the vale and is not bounded by anything besides convenience of access; the market transcends cultural identities. Because the monthly market usually involves only the male heads of family or widows, it does not act on the population at large to generate a wider sense of community. And the fair itself occurs only once or twice a year. Rather than important nodes of regional solidarity, the fairs and markets are overlapping zones in which participation is based above all upon practical considerations.

Perhaps the most practical sense of cohesion in a vale is the fact that each vale, as a municipality, has a variety of official organs that meet on a township-wide basis. The two most important are the junta, with delegates from each village and the official brotherhood of herdsmen. The junta—under the present regime a rubber stamp body for the provincially appointed mayor—has been a deliberative body of some dignity in the past and has administered roadbuilding and the communally owned lands of the vale.

Yet in the vale, as in the case of the Montaña, the inhabitants of the Nansa valley system rarely sense themselves as having a true community of interest with those villages that are their nearest neighbors. In the long verse epic written when he left the valley to work in Costa Rica in 1926, Manuel Agüera Vedoya lamented that he was leaving his wife and children in Cosío, where she was a stranger. His wife was from Celis, and he was from Rozadío, both in the same vale as Cosío, yet she was considered a stranger there. Villages within a vale are just as likely to be in litigation over disputed boundaries as those in adjacent townships. While the vale is a convenient handle to use in categorizing a group of people, it has not, at least in the Nansa valley system, elicited an effective sense of identity.

In short, with the possible exception of Polaciones there does not exist in this valley a significant cultural unit larger than the village with which to identify. While the Montaña has its share of dialects and local customs, it does not have either a language or a true dialect. As an appendage of Castile, with its heart for so long over the mountains in Burgos, it has had little time to develop its own ways, and now, with the homogenization and Americanization of Spanish culture underway by television, it will not have a chance to. And geography has militated against a narrower identity like that of Liébana to the west or Polaciones to the north. The villages in this narrow valley lie spilled up on hillsides, reticent and touchy. The valley has no central place, no town, no capital. Its citizens are therefore villagers without a *pays*, in a crevice between several *pays* but belonging to none—none, that is, except the village.

V. Village and Parish

The villages of the upper Nansa valley are all nuclear settlements. Until the twentieth century the only dwelling set apart from the village center was an occasional tavern along a highway. Why there are such dense clusters of houses is not clear. Was it perhaps for safety from bandits or wild animals? Or were these planned communities, built by colonizers from down the valley? Whatever the origin, it is certain that of type of settlement plays a decisive role in the overwhelming importance the the village as a unit of activity. Caro Baroja's analysis of the sociocentrism of the Spanish village applies nowhere more appropriately than to these villages. Each village is a tribe. The sizes of the villages seem to vary according to the size of the village territory. The size of the village territory is approximately the amount of land within a reasonable distance from the settlement, but some villages have corners of mountains to themselves, large backlands that permit them to support more households, and these are the biggest villages—the ones with the largest territories: Tudanca, San Sabastian, Uznayo, and Carmona.

The spatial arrangements of the villages have certain features in common. The houses are densely packed together, generally on hillsides up from the flat valley bottom. In a number of cases (San Mamés, Tudanca, Sarceda, Cosío, Obeso, Cabrojo, Celis, Camijanes) the church is set off from the village it serves, often on a slight prominence, as if it were added after the village, or perhaps as if to maintain a respectful distance from the village. In almost all other cases the church is on the edge of the agglomeration, free from congestion at least on one side (Tresabuela, Santotís, San Sebastian, Rozadío, Bielba). I am incapable of deciding which is more likely: that the village preceded the church; that the church or holy place was the reason for choosing the village site in the first place; or that the village and church were built simultaneously. I am inclined to favor the idea that the villages preceded the churches, if only because the traditional Spanish pattern locates a church or cathedral in the center of the urbs. There is some evidence, however, to indicate that in the earliest centuries of their existence, the churches were built to double as refuges or even fortresses, which might explain their location on prominences.

Around each village is the land its inhabitants till. The village generally dominates the lower, flatter mies as if to protect it. Indeed, the universal situation of churches dominating the mies might indicate the use of church towers as lookout posts. In the 1753 report one lower village notes the employment by the concejo of a man to guard the corn against boars and bears, and boars are still common plunderers of the corn today.

Since each village possesses open grazing land in common, each village is said to have a territory, or *término*. Some hills between villages are shared between villages; this arrangement is termed a *comunidad*, an arrangement that can be terminated by either village at any time. If no comunidad exists, the boundary between terminos generally runs along the crest of hills between them.

The villages are governed by a council composed of all male vecinos. A vecino is a head of a household, and vecino rights are calculated as follows:

> a married male = one vecino
> an unmarried male, living alone = one vecino
> a widowed male = one vecino
> a widow with children = one vecino
> a widow living alone = ½ vecino
> an unmarried woman living alone = ¼ vecino

Each vecino is granted equal rights to all communal property in the village. The council must make its decisions unanimously in the disposition and use of communal property. Its administrative board, currently appointed by the mayor of the vale, consists of a president, a secretary, and two aldermen. These gentlemen expedite the affairs of the village subject to the approval of the council. As far as the men are concerned, there is a great emphasis upon absolute equality in the council. Each has a right to his say, and the meetings can be very heated. Indeed, the pride and the fierce independence of these mountaineers is reminiscent of that of the Icelandic freeholders at the Thing, as described in sagas. The council may have come from the North with the visigoths; like the Thing, it involves a tension that often seems likely to break into battling, and which sometimes does.

The council will necessarily meet several times a year in every village. It alone decides whether and when cut fields will be opened to cattle, when the village roads will be repaired, and when the communal meadows will be cut or auctioned off.

The absolute equality within the council is echoed in the customs governing the distribution of property. All land and goods are divided equally among all the children, including daughters. It can be seen also in the uniformity of most outward habits of consumption of the villagers. In these upper villages, a man with seventy cattle lives to all outward appearances in precisely the same fashion as the man who has eight. With the exception of some noblemen and returning emigrants, inconspicuous consumption is the rule.

The council is the legal body of the village; the parish is the religious body. The mass, every Sunday, brings virtually every man, woman, and child into the church, all dressed slightly better than on work days. The seating divisions in mass reflect the divisions found throughout the everyday life of the village: Children are together in the front; next come the women over the age of attending school; finally, behind an aisle, the men—the professionals in the front, the old men in the back, and the youths and young married men standing in the balcony. Before and after mass the men stand and talk in rather stiff circles outside the church. They enter at the last moment and leave as soon as mass has been said. The women pass in past the men, and pass out through them. Hence every Sunday the villagers see their village as a social whole, and it has real meaning as a group of persons actively engaged together. The mass itself is set up as a common enterprise. Most of the village is there on the anniversary of the death of one of its members, rogations are said every year for the success of the crops, and various church festivals commemorate the different aspects of village life. The parish and the village are thus inextricably linked. From this perspective the difficulty of trying to turn a parish in a city or in a multireligious society into a community is most obvious. For the very boundedness of the village reinforces the sense of religious unity as nothing else could. A large urban church with masses every half hour has lost the essence of the parish, a sense of a community in worship before God.

In addition to the council and the parish, what other activities bind the entire village together, help form and maintain the sense of village identity? In these villages the entire work cycle ensures, at least, that the villagers will work together. The reason is simple and has already been adumbrated in the introduction. While all fields, both those in the mies and the private pastures, are walled or fenced off from the open grazing land, the strips within the fields and the mies belonging to each farmer are not. A map of these plots, if one existed, would show a quilt of little strips, the effect of endless subdivisions of property after inheritance. But even after the hay and corn is cut the land is valuable for grazing. In fact, by the time the fields have been cut the cattle and sheep need to graze on them, because the upper pastures have been overgrazed. This makes convenient the progressive opening of more and more fields to grazing animals as the summer goes on, which entails a coordination of the village haying so that the haycrop from the different sectors of the backlands is cut at the same time. Hence each village evolves a pattern or rough schedule for cutting: The Ribera might be cut first, that is, all those with land in the Ribera would cut it first; then the Sierra Yero, et cetera. So the effect is that at any given time in the summer, numbers of villagers are likely to be haying on adjacent fields. This in turn means that virtually all agricultural activities from ploughing and sowing the mies to harvesting it are done virtually simultaneously in the villages.

In several villages, the climax of the summer comes with the cutting of the communal meadow. In a vast and impressive ceremony in the communal meadow of Tudanca, all the mowers in the village gather around as their names are drawn from a bag and they are assigned lots in the order of the drawing. When the drawing is over they stretch out across the mountaintop field, a line of seventy men with long scythes cutting wide swathes up the steep slope. In order to achieve this degree of coordination, the village virtually has to work as a unit;

it starts harvesting on the same day, and it has to take care of the shorthanded and slow by giving them a hand so that everyone finishes up even and the animals can be let in to graze on more and more fields.

The degree to which each village forms a unit can best be seen when occasions arise for groups of villagers to venture away from home. At the fiestas up and down the valley the youth of each village cluster together. Thirty years ago this inevitably led to brawls between boys of different villages, and certain villages were known for their belligerence and bravado. Although things are tamer now, the village youth still travel in groups.

The same is true at markets and fairs. Just as the cattle of different villages are grouped separately at the market, so are their owners separate. At the San Miguel fair, which is held on flat ground below a hillside by the highway, the villagers from the upper valley arranged themselves in village clusters on the hillside for their picnic lunches, looking for all the world like their villages set up away from the road.

Because each village keeps to itself and forms a social world apart, each takes on a different accent in its speech, which even a relative novice can distinguish in time. In many villages there are special measures of land and produce used only in that village. Each village also has a nickname imposed by the other villages on its inhabitants, and each has its special peculiarity attributed by others. In this sense each village does indeed have its own culture, a culture that is accredited by the surrounding villages and validated by tradition.

While each village has a culture it can also be said that each village has a personality and carries a mood with it. The differences in atmosphere from village to village are almost palpable. They seem to be very complex in origin, but the following are some of the factors that affect them: the particular shape of the village's population pyramid of the given epoch; the distance of the fields from the village; the availability of grazing land; and the value of the communal meadow are all basic variables for different degrees of village harmony or, in local terms, of *unión*. Some of these factors would also affect another way the villages differ—their openness to outsiders. In the case of Tudanca, it seems clear that its work unity lies in the communal meadow and the fact that all labors have to be synchronized because of it. Phrases like "no salen hoy" (they are not going out today) describe what everyone is doing, for everyone or no one does things.

A key role must be played by the personality mix of the moment, the particular character of its Presidente and its priest, as well as the structural factors. I came to wonder, for instance, how much the presence of one genial anthropologist might contribute to the union, as an object around whom might form a shared village view of itself as hospitable, open, and loving. Just as the village unit congeals, in spite of internal disputes, when villagers are together away from home, so the villagers' sense of corporateness becomes clear when the village is penetrated by outsiders. The longer I stayed in any one village, the more evident it was that I was being reformed to fit the village—the conversion of an American to a Tudancu or a Bastianu.

> "Which do you like better, Spain or America? Tudanca or San Sebastian?"
> "You will have to marry a nice girl here."
> "You should ask for a share in the prado concejo."
> "Why have you deserted us to go to Obeso?"

The same gestation took place around a shepherd from another province who sold his labors to different villages for the communal flock. He was always being asked which village he liked the best, which people were most friendly. Both of us were called upon to confirm the existence and the quality of the village identity.

Perhaps for this reason a complete gestation never takes place. An outsider always has more usefulness to the villagers as a friendly outsider than as a completely converted insider. Hence foreigners, people from other places, who have married into the village are identified by their village or province of origin. In Tudanca lived "El Gallego" (even though he had been in Tudanca for 30 years), in La Lastra lived "Burgos," and in San Sebastian "El Sevillano." Rights to communal lands are granted to such persons only by the unanimous approval of the council, and years later it is not unusual to find such rights begrudged them by their enemies of the moment. These villages are very much closed corporations, and necessarily so, for the amount of wealth the land will produce is limited, and the inhabitants cannot afford to dilute it.

The necessity of keeping the wealth, coupled with the convenience of regrouping and reorganizing periodically the disastrously divided fields, is at the source of the endogamy of the villages, an endogamy that has broken down only in the last generation (see Table 3). As in many parts of Europe, the marriage of a village girl to a boy from beyond the village was, until recently, symbolically held up by the village youths, who demanded that the bridegroom pay a token brideprice to the village.

A critical measure of the existence of bounded communities lies in the extent to which such communities have a lexicon of their own. While each village has its own terms for certain technical matters, such as the name it gives for different sizes of heaps of hay and the name it gives to the different stages of immature cattle, each has two complete lexicons that are all its own. These are place names and nicknames.

When I moved to these villages, I found that one or two words in a sentence were often keeping me from catching the drift of conversations. As I did more and more work in the backlands around the villages, it often turned out that these words were place names. Each field, each barn, each path, each resting place, each prominent stone, each knoll, each spring, and of course, each peak and saddle in the village territory has a name. A survey of such names in one locality in Asturias turned up 600 names.[8] The names are by no means secret. Some of them even appear (to the villagers' great delight) on the military survey maps. But because people from outside the village have no call and no particular desire to learn all these names and places, the names become a kind of arcana, as personal to the villagers as the places and the land itself. When I set to learning these names, my growing fluency in local topography would be greeted with such exclamations as "Look what a Tudancu he has become already!" Just as the farm in New Hampshire where I am writing this study has its "north quarter," its "rabbit pasture," "blackberry hill," and "goose gate," so every conceivable landmark is catalogued in these villages, and the limits of this knowledge stop abruptly at the village membership, with the exception of those outsiders who cultivate land in the village that they gained through marriage.

The second lexicon that stumped me in conversations was the lexicon of nicknames. For a fuller understanding of the origin of the set of nicknames that parallels the formal names of people in the village, we turn to the whole institution of the *mocedad*, the corporation of youth in the village that seems to be a generating source for much of the sense of village solidarity in the adults. This same institution in the old days enforced the traditional endogamy of the villages.

As soon as children become ambulant in these villages they set up subversive friendships and play groups which undermine the rigid pattern of friendship and association of their parents. Their domain is the streets, the alleys, the porch of the village store, the bowling green, and the hills above the village. Like unseen matter the children dwell in the interstices of the villages, a world with its own geography and its own special places. Like the very old, the children are on the margins of society, and except for school and catechism are left to their own devices until they are old enough to be useful; then they are progressively admitted to the adult world, the adult geography, the adult special places.

Figure 2 roughly lists the age groups I observed in the village of Tudanca. The total number of persons in these categories is about 80. The play groups, for that is above all what they are, are conditioned and delimited by the institutions of the village: the division between crios and chiquillos (or niños) is whether or not they go to school; that between chiquillos and chavales is school—leaving age; that between chavales and chicos (or mozos) is whether or not they are allowed by their parents to go to dances in other villages, and in olden times, whether they had been registered for the compulsory military service. Yet although these rough limits are set by the village elders and the larger society, they are enforced and maintained by the youth themselves.

Both in San Sebastian and Tudanca Sunday afternoon games like hide-and-go-seek and red rover provided a good opportunity to see the nuances of age-grading. Those playing would make restrictions on who could join, like "no chiquillos," indicating four or five children hovering around who were on the

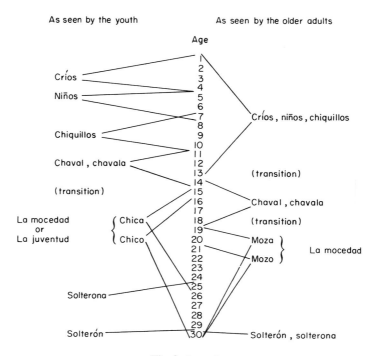

Fig. 2. Age sets.

lower limit of the age set. The clearest example of this kind of discrimination came when my presence in the village of Tudanca provided the *mocedad* with the opportunity to revive an old custom: the demand for a section of the communal meadow by the village youth. Because I was a foreigner and a guest in the village there was a better chance that the village men would grant the youth a section of the field (called a braña) if I asked for it, so several youths approached me on the matter. Fortified with brandy I asked for the braña in the council, and the vecinos graciously acceded. When the time came to cut and load the hay, there was a question as to whether the chavales were eligible to participate. It was their older brothers and sisters who told them that they were not. We auctioned off the hay we harvested, and planned an excursion to the city of Oviedo with the proceeds. When it was time to organize the trip, the limits of the age set became even more clear. In work groups and in small conversations it was agreed by consensus who was old enough and who was not, the line being drawn around age fourteen. Although we made it clear that all who helped work the hay were welcome on the trip, and several of the older unmarried men and women who had helped had said they were coming, they did not, in fact, show up at the chartered bus on the highway at six in the morning.

There was also some question as to whether the young married couples in the village could come on the trip. For instance, I was approached at the fair of San Miguel by two of the younger wives who wondered if a special invitation to them and their husbands couldn't be arranged. This idea was vigorously vetoed by the unmarried youth; marriage clearly meant an exit from the group. For Tudanca in 1969, the mocedad seemed to consist of unmarried women 15-25 and unmarried men 16-30.[9]

The same enterprise provided a sense of the mocedad as a unit erected in opposition to the adult world. Throughout the entire project, which ran over a period of a week from the first cutting of our hay to the trip, it was clear that we had to provide a united front to the village. The night we brought our hay to the village on sleds to auction off, the women smoked cigarettes and sang songs about the episode that could be heard in two or three other villages across the valley. In their songs they referred to me as the Presidente de los mozos, a term revealing that the mozos see themselves as a kind of youth equivalent to the village council. Finally, when the trip itself had ended (it was a high-spirited 24-hour excursion that had as its only conceivable touristic highlight the feeding of candy bars to Petra, a bear kept in a cage in the central park of Oviedo) there was a tacit conspiracy among the youths not to reveal to the adults what wild deeds had been done in the city on our day of liberation.

But it *was* fun, and we sang a lot of songs and told dirty jokes. They said an irreverent version of the Lord's Prayer, insulted people that the bus passed, all in all violating many of the village's unspoken rules. Again, the women all smoked. And it was all done, with the exception of a couple of minor lapses, in an open, free spirit. As the culmination of the play groups of childhood, the years of the mocedad are the high points in a villager's life, a time of maximal independence and minimal responsibility and worry. Marriage after this, especially for the women, is almost always a let-down; the establishment of a working cabaña is more work and more worry; and both marriage and the direction of a working unit are inevitable introductions to the competition, disputes, and drudgery, that accompany adulthood in the village. The mocedad, because they are not vecinos, are relatively free from the quarrels of their parents; children of parents who are not on speaking terms can be good friends. By this means the intensity of divisions within the village and their permanence from generation to generation are somewhat mitigated. Because of this exemption, the mocedad is the ideal group to organize the village fiestas.

Theirs is the golden moment of license—the role of insulting everyone in *carnaval,* the singing of *marzos* and verses under the balconies in the evening, and theirs is the opportunity for boisterous excursions to the taverns and dances in other villages that provide the battleground for the defense of the villages honor. The only requirement for membership in the mocedad, in addition to a requisite maturity and the unmarried state, is membership in the village; the mocedad is another institution, like the council and the parish, which depends on and reinforces village solidarity.

It is in the schools and in the period of the mocedad that the double system of nicknames is established. On the one hand there is a given name, some variety or abbreviation of which is utilized throughout life, but along side this name is often a sobriquet. This is characteristically based on a physical characteristic (a large nose, a small stature, red hair); some characteristic habit or tic; the result of an extraordinary event in the person's life (falling into a muddy pond); or the same nickname inherited from the person's father or mother. Such names are usually derogatory in a mild way, and so are never used in front of the person's face, but almost always used when the person is not around. They are usually not used maliciously, however, and only among persons who know them; usually only among those of the village. These nicknames are used exactly in the same spirit as collective nicknames for villagers are used by people in other villages.

Nicknames are created during the period of the mocedad because that is a time of very intense interaction when, in a spirit of fun, the different traits of the different members come to the fore. Each person has, in spite of himself, a public personality, one that is sometimes in sharp contrast to the personality advanced within the family. The period of the mocedad, then, engenders a set of characters or roles that are often maintained throughout the lifetime in the village, a kind of public theater; the roles are consecrated in the nicknames.

The maintenance of the roles and nicknames is made easier because the male part of the mocedad also continues. After marriage the men meet in the taverns, or in the fields, or at fairs and markets. Their chief interests are in their animals, and they meet to compare notes. The circle of older men is not so close as in the mocedad, for with responsibilities come enmities. But whereas the women stay in the home, in small sewing circles, or with relatives, the men maintain to some degree the freedom of association of the youth group.

It is the aspect of the village as a stage, with its own characters and its own standing jokes, that every day solidifies the village, that is the living reality of this nebulous thing known as community. People's public personalities get away from them; they become stock figures just as well known and appreciated after their deaths as characters in the Comedia del Arte, remembered as the essential elements of scenes and incidents recounted around the hearth, in the tavern, at the portal of the church, or in the winter barns, in the long chains of stories and anecdotes that form the village's cultural legacy and the village entertainment.

Indeed, this is one key as to why living in the village can be so engaging (pleasantly or unpleasantly); how the village comes to be a total institution providing a full gamut of emotional satisfaction. Once personalities are typified by nicknames they are to a certain extent domesticated. Women and men like to be called by their given name, not their nickname, and I sense here partly a fear of losing control. They want to be accepted on their own terms, not the terms that others (besides their parents) set for them. The others, on the other hand, enjoy capturing a villager, making him or her a lovingly *known* object. Once the elements in the drama are established, the drama itself comes in the manipulation of the elements, just as the plays of the Comedia del Arte or the guignols sometimes seen in the cities are basically recombinations of the same elements. The pleasure of watching comes partly in recognition, a found recognition of the favorite characters. The excitement when watching a village drama for the first time comes not in the novelty of the characters, but in the novelty of their intersections. The stories or incidents recounted in the fireside chains are the particularly felicitous recombinations of the known elements.

Possessions and skills become facets for characters in the village theater, but they are self-chosen, a way that people can regain a measure of control over their public roles. Because the village is a group of people, many of whom are doing the same things at the same time every year, they develop unspoken competitions and acknowledged experts. First of all is wealth, as measured by the number of cattle. While a cattle herd is unobtrusive, usually high in the hills, all villagers are aware of its magnitude. At the beginning of this century there would generally be a very few herds of any size belonging to individuals; the rest of the villagers would be divided between those who owned a few cattle and those with none. Over the century the wealth has evened out, probably because

TABLE 1

Animal Wealth: Tudanca, 1969

No. of cows in herd[a]	No. of herdsmen	Percent of herdsmen	Percent of total cows
70-75	2	3.8%	10.5%
60-69	0	0	0
50-59	1	1.9	3.8
40-49	7	12.9	21.9
30-39	7	12.9	17.3
20-29	14	25.9	24.9
10-19	18	33.3	18.0
0-9	5	9.3	2.8
	54	100.0%	99.2%

[a]Horses were counted as 2/3 cow; sheep were counted as 1/10 cow. The average size herd is 25 cows. The total number of livestock owned by the villagers is 1,305 cows, 127 horses, and 687 sheep. The few livestock belonging to those whose primary occupation is not herding were not included in the table. (These persons are the storekeeper, the electricity manager, and a number of elderly retired farmers.)

of the increase on the importance of the earnings of men and women in the off-season. At present the system is most clearly related to the number of active workers in the family unit, which determines the amount of hay that can be cut, which determines the number of cattle. More hands also bring back more money from the winter months. Table 1 shows the distribution of cattle and livestock capital in Tudanca.

In addition to personal evaluation by the number of cattle owned, there are acknowledged experts in many of the ancillary arts: cutting hay, sharpening scythes, the making of wooden shoes, laying stones for roadbeds during the communal labor gangs, grace in impromptu singing at fiestas, wit in the telling of stories, and the ability to compose the verse epics of the valley, the *trovas*. While these are masculine arts, there are also feminine arts. The most notable is beauty. Each village has its own informal contest, and in the old days the most beautiful girl was said to "llevar la gala" (be outstanding in quality)—the same expression used for the informal award granted to the herdsman whose cows came back from the summer pasture looking the best. Other feminine arts include singing, composing, cooking, and religious virtuosity. But the competitions are primarily carried on by the men. The men in a sense are the delegates of the work units to the village-wide competitions.

The village theater is facilitated by several conditions:

(a) The society must be small enough so that each person's personality can be appreciated for its distinctive traits, or at least small enough for a large enough number of local people to be known well enough to form a reasonable cast of characters. In urban societies or mass societies, where the characters in the cast are national figures—politicians, sporting figures, movie stars, royalty, or rock singers—the theater no longer reinforces a sense of local community. It is, in fact, perpetually transcending it.

(b) The society must have continuity. In an impermanent society the members are not around long enough to be typified, and a repertoire cannot be established.

(c) There must be a certain amount of leisure. The mountain society has much free time despite its hard work and poverty—the rainy days during haying season and the long winter evenings when the theater can be exercised.

(d) But most of all, the society must be focussed, must provide a stage. The settlement pattern is critical. It is difficult to see how such a complete theater could develop with a dispersed settlement. The theater comes from the juxtaposition of different personalities. It depends upon a high frequency and high intensity of interaction. It is the very essence—both source and expression—of villagehood.

VI. Barriada

The barriada is a subset of the village, containing three to ten vecinos. All these villages are divided somewhat aimlessly into courtyards, and some of them lie in two or three different flat places, as if spilling off one shelf onto another down the valley slope. The courtyards and hamlets provide the basis for smaller concentrations of activity, although the social groupings they engender are rarely formal entities and never as influential in an individual's life as the village on the larger level and the family on the smaller.

The courtyards are neighborhoods, and one naturally knows one's neighbors well, though not necessarily cordially. In one courtyard of Tudanca, known as Miranda, there are eight households. Three happen to be younger couples, who are good friends. Of these, one couple lives on a corner flanked by older relatives. Another old family is especially friendly with an elderly lady who lives alone across the courtyard from them.

Within these subgroups of Miranda there is considerable borrowing of tools and food and frequent visiting. The corner where the related families live forms the nucleus for a tertulia, or casual meeting place. These neighborhood groups are especially important for the very old and the very young. These two age sets, finding it hard to move around, have to treat with those physically immediate. For the rest of the barriada residents, associations with relatives in other parts of the village or activities centering on the taverns have more importance than those of the courtyard. While each location has a name, the group of people that lives there does not. It would rarely be said, "los de Miranda," for instance. At least not now. Older men recall intravillage disputes, with two gangs of youths which they say might have been based on two sections of the village. But that is something of the past, and even in the past, rivalry and fighting seems to have been far more between village bands than within them. Pereda's novels, which refer to these villages in the 1880's, bear this out. Although in *El Sabor de la Tierruca* he says, "Catalina salía con le gente de su barriada," it is the one single reference to intravillage subgroups, whereas a whole chapter, "Greigos y Troianos," is devoted to a battle between the youth of two different villages.[10] Of course, these villages have many intravillage disputes, but they are not, as a rule, geographically based, having more to do with kinship. There is none of the Mediterranean tradition of warring factions of upper and lower towns, for instance.

Another kind of casual social grouping without permanence or a sense of group identity is that which forms among persons with neighboring fields in a meadow around a set of winter barns. While each particular field is named, so are sets of fields. And in the summer, the persons working in sets of adjacent fields again form a kind of neighborhood. The shepherds and mowers know each other particularly well, visit with each other and help each other, loaning tools, hay, and labor freely. But neither the barriada not the prado has any ceremony nor any markers which set their boundaries.

VII. The Family Home

The final, most important social unit geographically based, the only one that transcends the village in importance, is the family home. It is in the family that the most intimate identity is established and maintained, an identity often in marked contrast to the role in the village theater imposed by the age set and adopted by the rest of the village. The name used for a person at home is the given name or a variation. The child, the husband, or the grandparent may not necessarily feel most at ease at home, but the home is a sure retreat. The nation, the region, the vale, the village, and the barriada provide increasing layers of ease and safety. The home is the ultimate refuge.

As with all the other identities we have discussed, the home is based on a place: the house where the family lives and where its ceremonies are held. Here, until recently, birth and death most frequently occurred. The house itself, like the village, is not a closed center: it is open to friends and relatives; but it is a secure place, relatively free from outside agitation. Quarrels between houses take place in the tavern or in the alleys and courtyards. The most frequent visitors to the house are members of the immediate families of the heads of household: parents or siblings, nieces and nephews. In these villages there is a fairly diluted, but nevertheless real, extended family system which serves as a mutual aid association in work, in the care of children, and in entertainment.

Besides a living unit, the family is also a work unit, a very organized enterprise, but whose composition varies considerably from household to household. We shall trace here the basic cycle of a family, first in a year of work, and then in the aging and succession of its members.

The family is the basic work unit, the number of able bodied persons needed to maintain a herd of cattle. The bare minimum is three persons. Single persons and young couples without help make alliances with others to get through the year. A work unit can be formed by any number of combinations: a young couple and an active father; a middle-aged couple with several children, some of them older than eight or nine; an old couple with teenage sons or daughters; or two unmarried older brothers or sisters. Every village has each of these combinations, but the herding enterprise has evolved around the nuclear family unit and the above types are substitutions for this unit.

In a year of work the family normally disperses and comes together several times, depending on the number of hands needed. The work it performs is fairly rigidly divided by sex and age. When the corn is planted in the spring, normally in early May, the father does the ploughing with horses, oxen, or cows, and the mother tends the corn, hoeing it at least twice during the months of its growth. Generally, control of the cows is in the hands of the father, with aid from the children, while the mother, again aided by the children, has charge of the animals at the stable in the village (the pig, the chickens, and the sheep) and the house itself. Since the village stable is either the ground floor of the house itself or next door, the mother's radius of action is limited, necessarily so, because she must watch the youngest children while she works. The father maintains the barns. The job of distributing the winter manure over the fields in spring falls to the children or the wife, aided by a burro, more often than to the husband.

As soon as the first hay is ready to be cut the family is brought up to full complement. Older children are taken out of the village elementary school, teenagers are brought back from work away from home, and those men working away return. During haying the men do the mowing, the women the raking, and with the exception of two villages, the men carry the hay on their backs and load the hay sleds and wagons. Children help with the raking, and carry hot meals from home to the hayers. The family continues reconstituted through the haying until the fall, when the supernumeraries again leave—the boys and younger husbands to work as woodcutters in village teams throughout northern Spain and the girls to work as maids in the cities of Torrelavega, Santander, and Bilbao. During the winter only two, perhaps one and a half, able bodies are needed to maintain the herd. One active soul is needed to feed the cattle morning and evening in the winter barns, and the other—a grandmother will do—is needed at home to cook for the children and maintain the domestic animals.

Sometime during the year, generally in late summer or early fall, the village fiesta is held, and all relatives within striking distance, including brothers and sisters living permanently away, come back and stay for a few days in the village. The climax of the family year is the feast held for the entire extended family in the home where the grandparent lives, after high mass on the fiesta day. For this meal a lamb is killed. The family tells stories, recalls old village songs they sang in the years of the mocedad, and individuals may get up and dance the jota.

The result of the division of labor in the family is that throughout the village there are some activities performed almost exclusively by men and others performed exclusively by women (see Table 2). This means that there are places in the village and the countryside that are the purview of one or another of the sexes. Men have the more honored tasks and positions. (We will discuss the duties of women as the family representatives to the Church and to God in the third chapter.) Because of this it is primarily the men who engage in competition over the cattle, and the women are to some degree freed of this emotional burden. The division of labor is even taught in the schools. On a scrap of paper I picked up in an alley of San Sebastian, the duties of the father, the mother, and the children are listed (see Fig. 3).

The older women tell of having been terribly oppressed by their husbands. In a prayer book, republished in 1946, that I found in one house in which I stayed, the following rules are given as guides for the behavior of wives, guides to consider when making their confessions:[11]

1. Esteem your husband
2. Respect him as your leader
3. Obey him as your superior
4. Reply to him with humility
5. Assist him with diligence

TABLE 2

Masculine-Feminine Division of Labor and Activity and Landscape

Division of labor/activity	Exclusively male	Predominantly male	50-50	Predominantly female	Exclusively female
	Plow	Cut corn	Load corn	Shuck corn	Clean house
	Govern	Mow	Tend sheep	Go to church	Rosary church
	Priests	Herd cattle*a	Spread dung	Tend mies	Cook in house
	Smoke	Sell cattle*a	Fiesta dance	Rake hay	Sunday communion
	Drink in bars	Carry Hay		Go to shrines	Tend garden
	Play cards	Drive cars		Get water	Care for children
	Play bowls	Ride horseback		Go on promesas	Buy, wash, iron clothes
	Village work crew			Remember old songs	
	Swear				Tend saints
	Professionals		Schoolteachers		Midwives
	Castrators			Tend pig, fowl	Witches
	Theatrical story-telling				Ring evening bell for animas
	Make wooden shoes				
	Write trovas				
	Take cattle to summer pasturage				
	Dance picayos (Obeso)		Dance picayos (San Sebastian)		Sing picayos
	Play bagpipe				Play tamborine
	Hunt				
	Tavern (night)	Winter barn	Alleys	Store (day)	
	Monte		Prado	Mies, garden	Wash house
	Council hall		School	Church	
	Market		Teleclubs set up by young priests		
	Bowls field	Stable	Bedroom	Kitchen	
		Portal	Bathroom	Balcony	

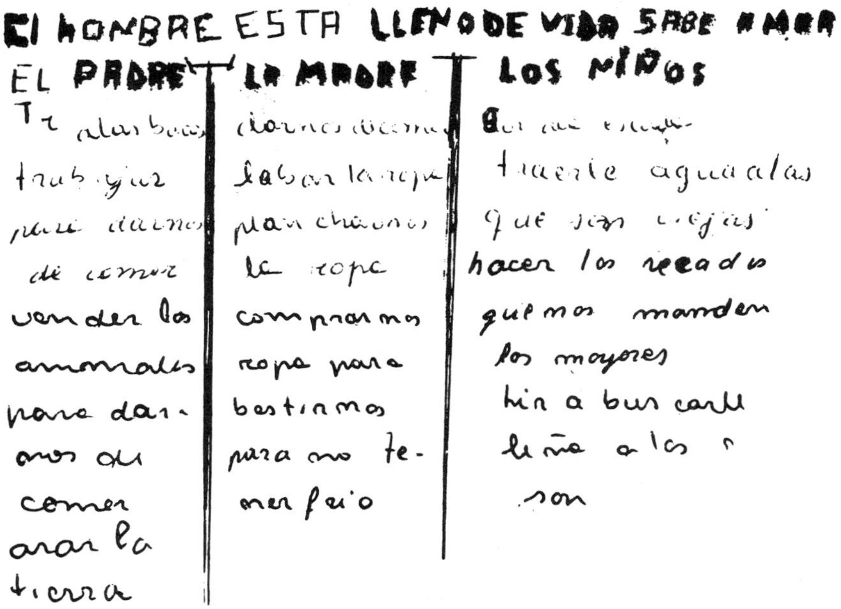

(translation of the duties of father, mother, children)

Man is full of life; he knows how to love

The Father	The Mother	The Children
Go to the cows	Give us food	Go to school
Work to give us food	Wash clothes	Bring water for
Sell the animals to	Iron clothes	those who are old
give us food	Buy clothes to dress	Do the errands that
Plough the earth	us so we won't be cold	older people order
		Go and hunt firewood

Fig. 3. Photocopy of a child's class notes, San Sebastian.

6. Help him with reverence
7. Keep quiet when he is angry, and as long as he is angry
8. Support his deficiencies with patience
9. Flee all familiarity
10. Cooperate with your husband in the education of his children
11. Do not waste the things of the house, or his wealth
12. Respect your parents-in-law as if they were your parents
13. Be humble with your sisters-in-law
14. Maintain good harmony with everyone in the house

The obligations of the husband include "direct her as your subordinate." The husband is not directed to be kind to *his* in-laws. Clearly the wife is seen as entering her husband's family. In *El Sabor de la Tierruca*, Pereda puts the theory of male supremacy starkly and succinctly:

And I believe, Nisco, that the marriage in which the husband does not know how to maintain his position is a bad marriage; and that position is maintained by the husband being worth more than the wife, that is to say, being lord and king of his house, not only because he is stronger, but also because he is more knowledgable in matters that concern them both. The more she has to learn from him, the more she will be proud of him, and the more she will be well thought of by the other people.[1] [2]

These are the old-time theories, and they are rarely put so blatantly today. But as to public behavior, they still apply. In public the woman is still expected to defer to her spouse and let him handle the honor and reputation of the family with the rest of the village, just as it is he who sells the cattle in market that they both worked to feed. The man gets the clothes dirty. The woman washes the clothes. There is an absolute double standard in regard to sex and entertainment. The woman is tied to the household; the man, in the course of his work in the countryside, at fairs, and during the months he works away from the valley, can get away from the village more easily. The council is all male, the priest is a man, all officials are males, and the ultimate legal power in the family rests with the husband.

The relations of women and men seem to be more nearly balanced in private than they were, say, 30 years ago. Within the household, a practical equality seems increasingly to be the rule. And the male dominance does not apply to children. In the children's games, a girl is as likely to order a boy around as vice versa. The change in public behavior comes during the *mocedad*, as males more and more take control of the group activities.

We have seen the household as a body that separates and coalesces during the agricultural year, that involves a certain hierarchy and division of labor. It also must be appreciated longitudinally as its members grow and die. As early as age five or six the child is able to help out, and by age 11 or 12 he or she is an earning member. The poorer families, in lean years, used to hire out their children at age nine as servants in houses in towns. Even now by age 11 or 12, spare children in large families are rented out by their parents for as much as one hundred pesetas ($1.40) per day, plus food. As the children grow older they are sent out to work as shepherds (age 14 and up), or loggers (age 17 and up), or servant girls (age 13 and up), during the off season, often earning enough to support the family during the winter months. Large families also pay off because in the critical months of haying, when every day of sunshine counts, the more hands there are mean the more hay can be cut, and the hay can be cut more quickly. So that, on the whole, large families have been the rule in these valleys: four to nine children is not unusual. The young couples by and large now practice some form of family limitation, because the old system of farming is less and less viable and the raising of children is more and more expensive.

As children reach the age of self-support, many of them go off to work semipermanently. As we have seen, three or four able bodied souls over the summer, two in the winter, are sufficient for the maintenance of the average-size herd—20 cattle—so that only two of the children are needed. Nor will the normal size holding support more than one or two of the children when the father dies. So it is decided who is most apt and has the most interest in taking over the herd, and who will seek their fortunes elsewhere. Those that leave trade their shares with their siblings for cash, or hold them to be worked by their siblings as an investment.

Spanish law requires that when both parents are dead, at least two-thirds of their land and goods must be divided evenly among the surviving children. In these villages the inheritance is almost always divided evenly. If children marry and set up households ahead of the death of their parents, it is common practice to give them their share of the land and cattle to start them on their way. Young couples work especially hard to buy more land and cattle in order to have a large enough herd to support children. This means that the young husband will spend as much time as possible away woodcutting or doing other seasonal labors to support the family through the year and thereby avoid selling the cows as they naturally multiply. In the meantime his brothers may be getting factory jobs in the city and his sisters may be marrying truck drivers, shopkeepers, or anthropologists. When the father dies, the widow stays on with unmarried children, if there are any left in the village, or if there are none, moves in with one of her married daughters.

If the division of property has not already taken place before her death, it takes place when she dies, and the old family as a unit comes to an end, splitting off into a new series of units. In fact when the fields must be divided, the herd split up, and everyone given a fair share, many siblings become enemies. Inheritance disputes are the single most frequent cause of estrangement and embitterment in the villages, because in practice it is very difficult to agree on who will have what. The division of property usually takes place by mutual agreement, but there is always the possibility that one of the siblings will take the case to court, which is considered a disgrace on the entire family. Using this as a threat, he or she can blackmail the rest of the family into agreeing to a settlement, and the rancor can last for years thereafter. The trials and tribulations involved in village-wide land redistributions, whether for convenience (*remembrement*, in France) or equality (as in China), are endemic within families in village societies where land must be divided equally among children.[13]

The raising of children, the focus of their affection, the person responsible for maintaining "good relations with everyone in the house," the tie that binds competing siblings, is the mother. The mother is the fulcrum of the family, the maintainer of the family home. The men go away and come back, either to the backlands to check on the cows or the coast to lumber. The women, their wives and mothers, stay, feeding the children and the livestock, until their return. The men have little to do with the children, to whom they are generally remote and austere. And, indeed, they have little to do with their wives. It is not uncommon when they are in the village to spend most evenings in the tavern watching television or playing cards. While the women are fairly cut off into the cells of homes, the men are able to maintain, if in a diluted form, some of the friendships and freedoms of the teenage peer group. For the children the woman epitomizes home, as for the adults the Virgin at the village shrine.

VIII. Relations With the Rest of the World: Aspects of Identities That Have Changed

From parish records and oral tradition it is possible to have a rough idea of the interaction between the upper Nansa valley and the rest of the world from about 1600 to the present. These sources indicate a fairly constant pattern of contacts between the valley and the outside from the earliest records until the last quarter of the nineteenth century, then some radical new interpenetrations of the valley with the outside world in the last 100 years.

In the process of searching the death registers of the parish of Obeso for mentions of devotion to shrines, I made note of mentions of Obeso people living or dying away from home. The records show that during the eighteenth and nineteenth centuries, men from Obeso went as youths aged 14-16 to Andalusia, especially to Cadiz, for work.[14]

Cadiz was one of the first Spanish ports to recover from the depression of the seventeenth century. The municipal tax records take an upturn in 1690.[15] Thus it is not surprising that mentions of Obeso boys in that area begin about 1715. In the 105 years from 1715 to 1820 there are mentions of 29 men or boys in Cadiz, Jerez, Puerto Santa Maria, Sevilla, and Medina Sidonia. Indications are that there were many more. Of these 29 men, 24 are young and unmarried, and at most five were married. Communications with Andalusia seem to have been erratic, for news of deaths were sometimes entered in the death register after a delay of six months or a year from time of death. Local tradition has it that the boys went down to serve in cafes. These would presumably be the *tiendas de montañés* in Cadiz referred to by Pio Baroja in one of his novels, "part store and part tavern," as they still exist in the villages of the Montaña.[16] And I know one elderly gentleman from Tudanca who went down to Cadiz as a youth to help out in the store of one of his uncles. Some stayed on, coming home only to marry, while others returned to herding in the valley. It is possible that the connection with Andalusia dates from as far back as the thirteenth century, when in gratitude for the aid of Santander boats in the capture of Sevilla, King Ferdinand granted franchises for certain goods to men of the northern ports. As a result merchant colonies of men from the north were established on the southern tip of Andalusia, and by the call of kinship, young men went down from the northern villages. Cadiz in the eighteenth century was the center of Spain's trade with her Empire, and thus a sure spot in an otherwise erratic economy.

From the ports in Andalusia the men, most probably younger sons not needed to maintain the family herd at home, seem to have spread out to the New World. In the same Obeso death registers there are mentions of 17 Obeso sons living in *Las Indias*, as the New World was referred to. Of these at least six or seven had brothers or fathers in Andalusia, which is why I think the colonies in Andalusia and the New World are connected. Those from these valleys again seem to have been called to make their fortunes in the Indies by uncles, especially mothers' brothers. And although the intermediate link with Andalusia has all but died out, the practice continues today. Today from every village in the valley there are five or six youths in the New World, especially Mexico and Guatemala, mainly working for uncles. Some of them will stay there, coming back to visit and marry, and in turn call their sisters' children over to help them out. In every village there is, or has been, a village character *El Indiano*, the man who went to the New World and returned wealthy. As Manuel Llano points out, those who go and fail do not return.

The Indianos have left their mark on these villages by building large houses, or casonas, especially in Cosío, Rozadío, and Tudanca, and by supplying the funds to rebuild fallen churches (the shrine of Los Llanos in Obeso, 1708, 1798; the parish church of Sarceda, 1815; the parish church of San Mamés, 1798; and most recently, the parish church of Obeso, 1960). In San Sebastian there is a public fountain recently built for the village by an Indiano; the altar was restored from its civil war devastation by a family from San Sebastian in Mexico. Those that return to stay may acquire a large herd of cattle and hire servants to tend them, sometimes disequilibrating the structured equality of the villages, or they may retire to a life of leisure. Every summer a certain number come back in automobiles and stay with their relatives, enjoying the fiestas up and down the valley and placing their children in boarding schools in Santander or Madrid. For most Indianos the village in Spain remains home. The countries in the New World are still, in a cultural sense, colonies to them.

In addition to the semipermanent migrations to Andalusia and the New World there were less permanent relations with Castile. The Obeso death register for 1718 carries the following notation: "He died in León, sawing there, at the beginning of this year, according to the news brought by some residents of this valley who were there with him." There is another mention in 1746 of a man sawing in the Guadarrama mountains near Madrid, and several other mentions for Castile which might refer to the same occupation. It seems clear, then, that what was an activity at the beginning of this century—that of teams of men from these villages going over into Castile in the winter months to saw logs into boards—goes back at least to 1700, probably further. According to old men in Tudanca, who practiced the trade in their youth during the winter months, they would go off, married and unmarried, from village to village in Zaragoza, Burgos, León, Segovia, and Valladolid, using huge two-story saws. Hipólito Grande of Tudanca, now over 90, made enough money by this means that from scratch he was able to purchase over 30 different plots of land in the village and eventually maintain a sizable herd of cattle.

The logging is continued today (although there are no more sawyers), and the younger men go off in teams of four or five, usually from villages in the same valley if not from the same village. They are now cutting stands of timber planted in the 1930's and 1940's in provinces closer to home, and they are not away for so long.

While more permanent contacts resulted from the work in the Indies and in Andalusia, because the logging was of such a short-term nature it caused few ties with Castile. A number of marriages took place, but there is no way of knowing how many since those marriages are registered in the parishes of the wives, where they took place.

Other regular contacts with Castile are mentioned in the 1753 surveys made for the Marquis of Enseñada: the sale of oxen from these vales and of poles, wagons, and wagon wheels from Polaciones. At that time there were also carters who traded with Castile based in the villages of San Sebastian, Pesués, and Polaciones. These contacts certainly lasted well into the nineteenth century, as they are still referred to by the older men.

As for the current pattern of girls' working as servants in cities during the winter, it is difficult to ascertain how far back this goes. It was a common custom, at least at the turn of the century when the oldest women now living were young. Two factors lead to the suspicion that the pattern is not of great antiquity. First, there are virtually no mentions of girls or women dying away from home. Second, the rate of marriage of village girls with outsiders remained very low and restricted to neighboring villages until about 1875, when it expanded to include husbands from neighboring valleys, and 1920, when it expended to cover a range beyond western Santander. It seems likely, then, that girls did not work winters away from the village until the end of the last century.

As was pointed out, the jobs in the Indies and Andalusia were obtained through relatives, and those obtaining jobs relayed them on through the family. José Martinez Martinez of Tudanca, now 72, was one of six sons. Three stayed home to help their father on the farm, and three went off, one by one, to work for an uncle who was a storekeeper in Cadiz. When times were bad in 1926, Manuel Agüera Vedoya of Rozadío left his young wife with two children and went to Costa Rica where he worked for a rich and politically prominent uncle for five years, sending money home, until he finally returned with enough wealth to establish a sizable herd. The jobs woodcutting were either obtained on a journeyman basis, as was the case with sawyers, or through jobbers in the nearby towns of Cabezón and Torrelavega, as is the current custom with logging teams. The positions as serving girls and jobs for men as laborers on down-country farms seem to be obtained through the chance establishment of patron-client family ties. Jancina Gonzalez, age 71, says a wealthy man in Puentenansa acted as a patron for her family and always had one of her brothers or sisters working for him. Similarly, the late Agustín Narvaez of San Sebastian fought in the war with an officer who later went on to rise in wealth and position, eventually becoming a provincial governor. He has always had one of Narvaez' several sons work for him on his farm. The servant jobs in the city are handed over from sister to sister or cousin to sousin. Even within the villages if there is a powerful manor with a need for servants, patron-client relations are generally developed between families, not merely individuals. The client family becomes a kind of privileged quasikin, with access to the kitchen and privileges such as the taking of fruit from the garden. In return they show a certain loyalty to the patron house.

The patrons of families, often called upon as godfathers, were their chief recourse when faced with legal or medical problems that transcended the village in scope. If a son was needed at home from the army or a family member was imprisoned or was a victim of bureaucracy, the patron did what he could. The patron families still are used as employers, but for other problems the village priest is increasingly turned to. He is generally the one educated man in the more remote villages, and he is ready and willing to be of service.

Villagers have forever gone out from the village, but there have also been outsiders coming in. In the seventeenth and eighteenth centuries the vale was staffed with outside professionals. Until recently the villages saw a constant influx of beggars; persons too old or deformed to work and with no inheritance and no family had no choice. Manuel Llano's father, who was blind, was such a man. These beggars made the rounds of villages and slept in stables and chapels. There were gypsies too, although not as frequently as in the south of Spain.

In recent years two other influxes of outsiders have made their mark on the valley. The first were the road crews that built the highway at the turn of the century, several of whom married local girls. These men seem to have been mainly from the valley and from neighboring valleys. The second influx was of laborers on the Saltos del Nansa dam that was constructed in the 1940's between Polaciones and Tudanca. The construction of the dam brought a measure of prosperity to the upper valley in what was generally a severe depression. Coincidentally, it also emptied the villages of their unmarried girls. Now the dam and the power stations are manned both by villagers and outsiders, and they constitute tiny nucleae of modernity in the valley.

The rate of interaction with the outside world seems to correspond to periods of expansion and decline in the national economy. The death registers show a drop-off of contacts with Andalusia in the 19th century, coincident with the national economic debacle that followed the loss of the colonies. Similarly there are few notices of contacts with the South in the seventeenth century, which was also a period of economic depression. In contrast, the eighteenth and the twentieth centuries stand out as periods in which the villagers traveled far afield. This correspondence makes perfect sense, for in times of depression, jobs are scarce, food is expensive, and one is better off staying at home on the farm to eat. Similar findings exist for other village studies.[17]

One of the more recent changes is that the city is increasingly looked to as the ultimate destination. The better-off families consider selling the cows and buying an apartment. Commercial families will occasionally send their children to secondary schools. But the major change is the rate of marriage of the girls with outsiders. This has jumped dramatically in the past 30 years and continues at a high level. The endogamy of the villages, along with the entire village economy, is fast ending. Of the five villages whose marriage records I studied, four had a rate of less than 50% of marriages of partners from within the village. Fewer and fewer girls want to be farmers' wives. They yearn for a life of comparative leisure in the city. When they hire out as serving girls they are also looking for husbands.

While the rate and direction of contacts with the outside world have changed, the pattern remains fairly constant. The ties between the villagers and the outside are still mediated above all through the family: as a unit of patron-client relations it is a safeguard for all parties, for the patron is known and trusted through experience and the client is one of a family that already has been tested, a family that will answer for its member should he or she fail in any way. Because of its individualistic contacts with the outside, the family has always been in a certain tension with the village as a social unit. But in the past, the loyalty of family to village was strengthened by the family capital, the herd, being tied to the village through land and vecino rights, and the consequent advisability of alliance through marriage to other families in the village. Under those circumstances marriage with outsiders meant throwing away half of the resources of the prospective couple. Now that the herd is seen as an encumbrance, as capital to be liquidated, exogamy is no longer unreasonable. Now each family is for itself, and each family makes its peace with the outside world, placing its sons and daughters as advantageously as possible, this time not for the ultimate purpose of bringing some of them back to start life on their own, but rather as outposts of the family in the outside world, potential supporters of their parents in their old age (see Table 3).

Family cohesion, then, has not particularly suffered. It has adapted to new conditions. Lines of communication are more spread out than they have been since the eighteenth century: members are all over Spain, or at least the north of Spain, as a rule. What has suffered under the industrialization and urbanization of the Spanish economy has been the village as a social unit. Its magic leaks out year by year. As an ensemble it has less and less unity. Its ceremonies, its own

TABLE 3

Marriages: The Decline of Endogamy 1875 to Present[18]

Village Time period	Percentage of marriages			Number of marriages
	Within village and with neighboring villages %	With Saja and Deva valleys %	Outside region %	
San Sebastian				
1859-1874	91	9	–	32
1875-1899	98	2	–	41
1900-1919	88	12	–	42
1920-1939	75	6	19	32
1940-1952	79	–	21	33
Sarceda				
1852-1874	87	10	3	38
1875-1899	83	16	3	32
1900-1919	75	18	7	28
1920-1939	90	10	–	20
1940-1967	66	10	24	50
Tudanca, Santotís, La Lastra				
1725-1749	95	5	–	107
1750-1774	95	4	1	109
1775-1799	96	2	2	93
1800-1824	87	7	6	90
1825-1849	85	9	6	70
1850-1874	88	7	5	102
1875-1896	79	12	9	90
1897-1939	- - - - - - - - - - - - - - - not available - - - - - - - - - - - - - - -			
1940-1968	53	6	40	154

particular culture and personality, have become more and more embodied in its older residents. Like the farmers in Ploedemet and Chanzeaux, the herdsmen of the upper Nansa think that in 20 years it will all be over. With the city as the focus for attention, the village is without hope and its ceremonies empty. Its own peculiar theater has lost its audience. The only lively spirits at the fiestas are the youth, who will be leaving, and the Indianos and the city migrants, returning for summer vacations to the pastoral landscape of childhood memories. These migrants come back from time to time disenchanted with their urban experiences, but for the villagers the city is the place of enchantment. The younger priests, who expect an exodus, try to prepare the youth for city life. The older priests, who see a lifetime of trying to maintain village unity lost to despair, mark time at the card tables.

The television has greatly abetted the reorientation of families away from the village toward the city. With its pictures of an automated, mass-consumption life style, it has been an important source of dissatisfaction with the physical costs of rural living. While many of the programs are American, and hence classifiable as virtually from another planet, the advertisements, generally for goods the villagers cannot afford, are Spanish. No attention or value in the programming is placed upon a rural existence. As in America, on whose television Spanish television is based, the programming is the distilled essence of urban, mass-consumption living. As I said, the villagers know that the villages are at the end of the road, and while I believe that they have always known this, the constant, nightly reminder of an alternate way of life is devastating in its cumulative effect. Time and time again they approached me to provide confirmation of their own sense of relative ignorance and poverty in comparison with the rest of the Western World. It is this sense of sour dissatisfaction that makes the notion of the maintenance of the village community a chimera, that makes emigration inevitable, that makes the effective termination of these village societies a foregone conclusion. Psychologically, in a very few years, they have been wiped out. The economic forces that have provided higher standards of living elsewhere and then provided the means for the spread of this knowledge have done their work already.

What may eventually happen to these villages, since they are blessed with a setting so beautiful, is the kind of wholesale exchange in population that has taken place between New Hampshire and Vermont, on the one hand, and Boston and New York, on the other. The villagers will go to work and live in the cities and well-heeled disenchanted city dwellers will take over the villages. The signs of a nascent reoccurrence of the 1890 cult of rurality are already present in present-day Spanish cities.

To summarize—the individual in the upper Nansa valley has a series of identities of greater or lesser importance. Those most important are his village and his family. The former is gravely challenged by the relative prosperity of the urban culture. The latter is flexible enough to survive even when the village prosperity fails. In addition to these identities the person has a more diluted sense of membership in an age-set, in the vale, in the region, and in the nation. Largely due to increasing amounts of contact with people outside the village, the smaller units are losing ground to the larger units as foci for identity.

The employment away from the villages seems to be critical for the establishment and maintenance of the wider identities. The logging groups are often formed from men not just from one village, but from the vale or neighboring vales. And this helps to establish and maintain a sense that the larger valley is a social as well as geographical entity. Intermarriage comes into play here, for although the rate of marriages between vales has rarely risen to as much as 10%, even that is enough for there to be permanent links between families up and down the valley and visits made to the houses of relatives for the different village fiestas. Similarly, a sense of cohesion as Montañeses seems to result from the colonies of Montañeses in Andalusia and the Indies, and the Santander regiments in the army. The farther away from home, and the sparser the group from the home area, the wider the home area must be expanded to encompass enough kindred spirits for a little society. In the following chapter these identities are related to religious devotions. Devotions too are becoming less localized and more general in orientation.

CHAPTER TWO

THE SAINTS: SHRINES AND GENERALIZED DEVOTIONS

I. Introduction

Corresponding to all the geographical levels with which the inhabitants of the Upper Nansa Valley identify, there are specific divine figures to whom they can address their prayers, their offerings, and their thanks. The first part of this section will trace the advent, map the scope, and consider the role that the various figures have played for collectivities over the past 300 years.

The saints of importance to collectivities are statues holding fixed positions in the landscape, whether in chapels or in parish churches. There are many such devotions, spatially based, geographically located. In time each has come to have a territory of grace, an area over which its benevolent power seems especially manifest. Such images, even if in terms of name and design they are not unique in the vicinity, are considered particularly powerful. Henceforth we will call them *shrine images*. Their location, whether it be a chapel or a parish church, we call a *shrine*.

I presented the ethnography of the valley in terms of levels of geographic identity and the activities on which those identities were based for the precise reason that the territories of grace of the shrines throughout Spain are often these same geographical units. For the human community that lives in its territory of grace the shrine image comes to symbolize the landscape and activities on which the identity is based. The shrines are seen by the people in the territory of grace as assuring the successful prosecution of the vital activities of the group. It is a transaction point in the landscape between the human group, the land, and the powers that influence the success of the group's enterprises.

The metaphor that has found most favor in describing the relationship of shrine image to its territory of grace is that of patronage. As late as 1811 the nation, the regions, and the vales had a hierarchy of lords and overlords to whom certain taxes were due and from whom, at least theoretically, peace and protection could be expected. Similarly, the pantheon was seen as a court, with the Virgin Mother and saints as Queen and courtiers, intercessors with God, the King. While the earthly equivalent of the patron-client system has been somewhat transmogrified in the last 200 years (a more recent analogy made to me referred to the saints as like cabinet ministers and God as like Franco), the metaphor of patronage for the divine figures remains. Much of Spain is divided up and allotted to different divine figures as their special baliwick, or *patronazgo*. The division, while sometimes formally confirmed by the Bishop and the Vatican, is in principle predicated on the devotion of the people themselves, who legitimate the patronages by their allegiance.

Because the people, spread out unevenly across an uneven landscape, form clustered communities, valleys, and regions, so too the patronages follow these identities. Just as some of the identities are based on formal institutions (the nation, the province, the bishopric) so some of the patronages are formal patronages, created or publicized to provide the institution with a symbol and a divine guide. And just as other identities are almost exclusively cultural, based on a community of symbols or activities that crosscuts the administrative or institutional framework of the nation, so other patronages apply to a territory whose boundaries are exclusively cultural. There are other shrines whose territories of grace seems to be purely on the basis of access, like fairs or markets, and unrelated or only vaguely related to a community of activity or symbols among its devotees.

The reason that some shrines have a relatively bounded territory of grace and others do not is that the land and its cultures are not themselves always well divided into bounded cells. In some parts of the country, generally the more mountainous parts, valleys do form what have been called natural regions. But the maps that cultural geographers have made of the natural regions of Spain show less than half of the land surface falling into clearly defined natural regions.[19] And the whole matter is very nuanced; some regions are more bounded than others. Here is a clear divide formed by a mountain chain; there one district shades off into another. Hence it should not be surprising that some of the shrines whose devotional history is discussed in the first part of this chapter have clearly delineated purviews, and others do not. The social and cultural zone of the shrine is linked to the landscape, and its own degree of natural boundedness.

The first section considered the extent to which natural regions and common experiences have engendered group identities among the people who inhabit them. By classifying the shrines used by people of the Upper Nansa Valley according to the breadth of territory of grace, we will be able to see the extent to which there are shrines at the different levels of identity noted previously.

The second portion of this section deals with devotion not directed to shrines. Shrines as we have defined them are the homes of specific images that in a certain sense are not susceptible to substitution. They are specially charged transaction sites, with well-proven intermediaries for communication with God. They are implanted in the landscape. Some of the shrines have denominations that are unique, like Our Lady of Las Lindes. There is no other shrine of that name in Spain. Others have more common names, like Our Lady of Mt. Carmel. But if the image is considered special, if it has developed a territory of grace, then it is not merely Our Lady of Mt. Carmel, but Our Lady of Mt. Carmel of Cosío, or of Sopeña, or of Nava de Santullán. That is, it too becomes unique and is distinct from other images of Our Lady of Mt. Carmel. Its uniqueness is conferred upon it by its location. The location itself, impregnated by a tradition of worship, often confers upon a statue its territory of grace.

These unique devotions, these one-of-a-kind shrines, owe their uniqueness to their connection with the landscape. They are, as we have said, located images. Historically they have drawn support in their establishment and maintenance from the institutions located in that landscape: the Bishoprics, monasteries, Lords, and local governments. They seem to have existed from the inception of Christianity in the countryside, and there is evidence that some of the shrines predate Christianity. They contrast with the second type of devotion studied in this chapter, which could be termed *generalized devotions*. These are not shrines. They are devotions that are eminently substitutable, that is, if they involve an image or a painting, any similar image or painting will do. Their place in the local landscape is irrelevant. Persons relate to them as individuals, not as members of collectivities. The images are not considered to have territories of grace. The images themselves are not considered special. They are symbols, with no value in and of themselves. Most of the images in the parish churches are of this nature, as are most of the lithographs in the valley homes. These devotions, like, say, Our Lady of the Sorrows, have been propagated since the middle ages by the nonlocated institutions of the Church, those elements that were relatively unfixed or unconnected to the local landscape, such as the mendicant orders, the teaching orders, missionaries, and the Vatican itself. Their growth and spread corresponds to the growth of Christendom-wide institutions and nationwide institutions—the growth of Papal power by way of the religious orders over the power of the bishopric; the growth of national power by way of the centralized monarchy over the local lord. For these figures, like the local divinities, have been used as tools to manipulate the population, as well as incentives to stimulate a richer and more intense communion with God.

The distinction between these types of divine figures— those located in shrines and those whose location is irrelevant—will be emphasized by examining them and their histories separately, because in some ways the two types of sacred figures are essentially different. The shrine image seems to engage the social self of its devotees. They approach the shrine not only as individuals, but also as members of collectivities. The shrine ceremonies are social as well as religious statements, reaffirmations of identity and solidarity. For this reason they are supported by those institutions which are based in the territory of grace, institutions that depend for their survival or smooth functioning upon the sense of identity that the shrine devotion consolidates.

Aside from some of the brotherhoods, generalized devotions lack this component of reverence by individuals as members of larger groups. Instead they serve the individual *qua* individual or become the focus for family religion. Their propagation by the orders came for the purpose of encouraging a personal interest in redemption and salvation, which itself was seen as a personal or, at most, family enterprise. The two types of divine figures—shrine images and generalized devotions—represent two poles in the Church: the local Church and the universal Church. In fact it appears that until recently the constant attempt of the universal Church to integrate the local community into the community of Christendom by way of the generalized devotions was a continually losing battle, as the local communities would appropriate generalized devotions for local use as shrine images. Hence there has been a certain succession of shrine images and a constant influx of new generalized devotions. The recent increase in popularity of both generalized devotions and more regional shrines seems to parallel the shift in interest away from the villages noted at the conclusion of the previous section.

In these pages the word *devotion* is used in several senses. "Devotion" is the attitude of reverence toward a sacred figure. "A devotion" is an ongoing reverence or cult accorded a sacred figure (as in shrine devotions and generalized devotions). Some persons are more devoted to some sacred figures than others, but particular devotion to one sacred figure by no means rules out devotion to others. Most people in the valley are devoted to several divine figures. In a more limited sense, a devotion is also a formal act of worship, such as a prayer, a novena, or the stations of the Cross. The rules and formulas for such devotions are available to people in prayer books known as devotionals (devocionarios).

An advocation is a variant form of a divine figure. The Greek gods each had a number of advocations; each advocation corresponded to a different characteristic or to the location of a different temple. For each name there would be a different kind of image (perhaps differently dressed or carrying different implements) although the god remained the same. Similarly in Roman Catholicism the same divine figure can be known by different characteristics. Two examples for Spain are Saint James and Christ, which go under a variety of names, each with its corresponding form of statue: Saint James the pilgrim, Saint James the Moor slayer; the Child Jesus, Jesus the Nazarene, the Crucified Christ, the Christ of a specific place.

But the most common case is that of Mary, who has thousands of advocations, some standing for significant moments in her life (Immaculate Conception, Purification, The Sorrows, Assumption), others for the devotions sponsored by various orders (The Rosary, Mt. Carmel), and others, the largest number, for the sites of her shrine images. Most of these advocations will be recognized as versions of Mary because they are preceded with the words "Our Lady of . . . ," or "Saint Mary of . . . ," or "The Virgin of . . . ." As with the Greek gods, each advocation is related to the particular form in which Mary is represented. The images relating to her life or the generalized devotions like the Rosary each has a certain identifying characteristic, especially the color of the robes she wears. With the shrine images the distinctive form is the precise statue in question. When these images in turn become generalized devotions, as with Our Lady of Lourdes, Our Lady of Fátima, and Our Lady of Montserrat, the home statue is merely duplicated.

The Church maintains that there is only one Mary, that all representations of her are interchangeable. But devotions have crystallized around different representations, a manifestation of the inevitable problem of localization mentioned above. That this should be true in the case of Mary is particularly ironic, for it seems that originally, when the cult of Mary was propagated in western Europe by the mendicant orders, Mary was a symbol for the universal Church and her cult seemed to challenge the local cults that had developed around bishop-saints and other local favorites. The parishes, dioceses, religious communities, and nations (with specific shrine images) and eventually even most of the orders themselves (with typified generalized images) countered the universalistic impulse by diversifying the image of Mary. As a result, even though the same divine figure reigned everywhere, the symbols for different communities were distinctive, and each community could be said to have its own Mary.

The people of the Nansa valley both do and do not distinguish between the different varieties of Mary available to them. They all know that "Mary is one" in theory, and some of them make no distinctions in practice among the different advocations. For various reasons (see Chapter Three) many others, however, are more devoted in practice to some advocations, whether they be shrine images or generalized devotions, than to others.

II. The National Shrines

Spain has three official divine patrons: The Virgin of Pilar, St. Teresa, and St. James. Of these, only Pilar in Zaragoza has attracted any devotion, as a shrine, from the valley. The earliest evidence of such an interest is in the use of Pilar as a given name, in 1854, in two different villages. Since then there have been one or two women at most of the villages named Pilar at any given time. The devotion was probably brought to the village by the men, who in the course of their winter work as sawyers in Aragon or in military service visited the shrine. I know this to have been the case for Tudanca, in particular.

The Virgin of Pilar was a symbol for the nationalist cause in the Civil War, and in the north of Spain it is not unusual even today to see the lithographs of loyalist planes bombing the shrine that were distributed during the war to fan pronationalist sympathies. Pilar is the patroness of the Guardia Civil, and villages like Puentenansa that have Guardia Civil barracks mark October 12, the feast day of Pilar, with open house at the barracks and a dance. The schoolchildren in religion and history classes are taught the pious national legend that the Virgin of Pilar appeared to St. James at Zaragoza in 55 A.D. to reassure him as to his sacred mission of evangelization and to found the shrine with a sacred pillar. Yet there is no special devotion to Pilar aside from the women who bear her name. The shrine itself is to the Nansa villages what the Liberty Bell is to New Hampshire villages.

St. Teresa seems to have attained some popularity in the valley in the sixteenth and seventeenth centuries, which is when statues of her appeared in some of the valley churches, but there is no evidence of any interest in her shrine in the province of Salamanca. What little attention was accorded to St. Teresa was as generalized devotion. The same holds for St. James, although his fiesta is a national holiday. Even as a generalized devotion Santiago has never found favor in the valley. The name Santiago, as a given name at baptism, like that of Teresa has never been in vogue. And no one I spoke to in the valley had ever been to, or expressed any interest in going to, Compostela.

Other distant shrines that have a nationwide attraction are those of the Virgin of Montserrat, in Catalonia, and the Virgin of Guadalupe, in Extremadura. Montserrat has just begun to be used as a Christian name in the valley in the last 10 years. It too, like Pilar, exercises a certain tourist attraction, but virtually no devotion because of its distance. As in the case of Pilar, more valley people will probably visit Montserrat as automobiles become more common. (At present there is, on the average, only one car for every 20 households in the valley.)

The shrine of Guadalupe seems to have had some popularity in the seventeenth and eighteenth centuries. In 1768 a San Sebastian girl was christened Maria Guadalupe, and a shrine dedicated to Guadalupe in Treceño, 20 kilometers north, was mildly popular at this time. A Basque chronicler reports that at least as early as 1622 monks from Guadalupe circulated throughout Spain seeking alms and taking orders for masses to be said at their shrine.[20] And from the Guadalupe monastery records and annals it is clear that the eighteenth century was a time of unparalleled expansion and wealth. Perhaps the regular arrival of sheep and shepherds from Extremadura in Polaciones had something to do with the valley's flicker of interest in the shrine. At any rate that flicker does not seem to have survived the monastery's plunge of fortunes when the Hieronymites were expelled in 1835. At the present time, with the possible exception of Covadonga (discussed below) no national shrine attracts a steady devotion from the people in the valley for the simple reason that they are too far away. But the use of names from the national shrines as Christian names in most recent years shows that as more and more of Spain comes into the valley, the symbols and patrons of the rest of Spain gain in attractiveness. The increase of interest in national shrines parallels the intensification of national identity noted in the first chapter.

III. Regional Shrines

The shrines favored by the valley on a more regional level can be followed by studying the death registers of the parishes. Vital statistics began to be kept by the parish priests beginning about 1600 in the aftermath of the Council of Trent. The priests were given directives to pay special attention to the listing of masses to be said for the soul of the deceased according to his or her will, and in many parishes these masses and the shrines at which they were to be said were faithfully listed. The custom of having masses said at shrines after death seems to have been in full swing when the records began to be kept and continued until the nineteenth century. Hence for the period 1650-1820 the death registers provide the best information available about the relative popularity of the different shrines in and around the valley. The village of Obeso has the most complete records, but the fragmentary records of San Sebastian, Sarceda, and parishes in Polaciones confirm the general configuration of Obeso devotions, with variations only in the most local shrines. Figure 4 shows the relative frequency with which masses to be said at different shrines were ordered, over time, and Fig. 5 shows where the shrines are located in relation to Obeso and the rest of the valley.

Without exception the regional shrines have all been at one time or another monasteries or collegiatas. Monasteries frequently need and like to have

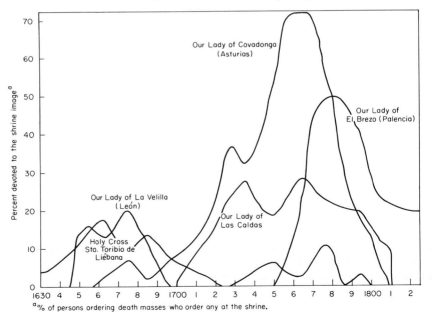

Fig. 4. Graph of Obeso devotion to regional shrines.

Fig. 5. Map of shrines frequented by persons from Obeso, 1600-1969.

a popular image in order to attract income, and conversely, a popular shrine demands the presence of clergy in constant attendance, especially if there are large numbers of memorial masses to be said. So there are cases of monasteries being set up on a site because a shrine is already there, and conversely, of monks propagandizing the powers of their patronal image throughout the surrounding countryside, thereby regionalizing what had been a small shrine or creating a new shrine from scratch.

Another common denominator of the regional shrines frequented by the Nansa valley people is that they are all located in areas primarily given over to herding. Thus they all share the same environment as the villagers, and the images worshipped could be considered to have a special interest in the kinds of problems the Nansa villagers might face. Only very rarely is a shrine listed that is in the wheat-producing areas of Castile, just over the mountains. In the Obeso records the Christ of Burgos, a famous shrine run by Augustinians in Burgos, is mentioned only once. The popular shrine of Nuestra Señora del Valle, at Saldaña, Palencia, is mentioned a couple of times in the Polaciones records, but this is because certain taxes from Polaciones had to be paid at Saldaña every year. The other contrasting ecological zone is that of the coast, to the north. Of the thousands of masses listed from Obeso in the 200 year period, there is only one mention of the noted shrine in San Vicente de la Barquera, Nuestra Señora de la Barquera, only 36 kilometers away. The shrines they do go to are in herding country, consistent with their one, diffuse regional identity—herders.

The order of discussion of the regional shrines will be that of their appearance in the records—the order of their historical popularity. Aside from the Franciscan monastery of San Vicente, which seems to have had special privileges for saving souls from purgatory (the testaments usually mention an *altar de gracia*) and which was not, properly speaking, a shrine, the first regional shrine to appear with any degree of frequency in the Obeso records is that of Our Lady of La Velilla, 123 kilometers away near the village of Mata de Monteagudo, León. The first mention is in 1645, the last in 1766, and most of the Obeso devotion seems to have been concentrated in the years 1656-1681. No one from the valley goes there today; no one has even heard of it. Devotion to La Velilla, the official patron of the district of Riaño, León, is now limited to a radius of about 15 kilometers around the shrine. The shrine's feast days are Easter Monday and October 5, when about 2000 people show up for the ceremonies.

La Velilla was a monastery during the early middle ages, when it was known as Santa Maria de Vallulis. The monastery was abandoned in the fourteenth century, and the shrine fell into ruins. Tradition has it that sometime in the fifteenth century an impoverished nobleman from the village nearby, collecting stones or clearing land, came upon the statue of the Virgin, took it home with him, and was punished for keeping the devotion private by a series of Jobian afflictions. His tribulations ceased only after he placed the statue in a small oratory on the old site for public veneration. After that he and his family flourished as never before.[21] Devotion to the image grew to large enough proportions by the late sixteenth century to make possible the construction in 1606 of the substantial chapel now on the site. The apogee of the shrine in terms of popularity and breadth of devotion zone appears to have been the seventeenth century.

La Velilla is the most distant shrine from the valley with any history of attraction. It lies beyond two mountain ranges, and to get there and back on foot must have meant a four or five day trip. Its fame may have spread to the Nansa valley by way of priests in the Diocese of León, nearby in Liébana. It has been maintained since its refoundation by chaplains.

After its initial flare of popularity La Velilla dropped out of the repertoire of the Nansa valley. The same thing has happened to the shrine of St. Toribio de Liébana just outside Potes, 44 kilometers to the west. At the monastery of St. Toribio is venerated what is supposed to be a fragment of the cross upon which Jesus was crucified. Known as the Lignum Crucis, the relic was supposedly brought from the Holy Land by St. Toribio to Astorga in the fifth century, and thence it found its way to Potes to be kept from the Moors. There is documentary evidence that the monastery dates back to the eighth century, and the Lignum Crucis was there at least as early as 1316. The death registers of Obeso show some masses in the first half of the seventeenth century and fewer in the eighteenth century. The last mention is in 1799. The records of baptismal names of San Sebastian confirm this timing. Toribio was used as a given name with some degree of frequency until the beginning of the eighteenth century.

The Benedictines of St. Toribio were exclaustrated in 1835, and only with the entry of the Franciscans in the twentieth century has devotion of any breadth resumed. But now, as in the case of La Velilla, the devotion to the Lignum Crucis is limited to the surrounding district, in this case the district of Liébana. There an active brotherhood is maintained, and the monastery is used for retreats and various civil and religious ceremonies. The only vale in the Nansa Valley that still maintains any important devotion to St. Toribio and the Lignum Crucis is Polaciones, which borders directly on Liébana. Persons from the other vales have been to the September 14 feast, but not, it seems, for religious reasons.

La Velilla and St. Toribio were attractive to the valley folk in the last half of the seventeenth century. The year 1700, however, seems to have marked a decline for a number of shrines, with new shrines succeeding them in the favor of the villagers. The first of these was Covadonga, the cave in Asturias whence the reconquest of Spain from the Moors is said to have begun. A number of shrines across the north of Spain have claimed to be intimately associated with the origins of the reconquest—cradle of the republic shrines, as it were. Covadonga is foremost among them, and its story, like Pilar and Santiago de Compostela, is told in the religion and history books used by the schoolchildren. In this cave Don Pelayo, refugee King of Toledo, fought an onslaught of Moors with a handful of soldiers. He later went on to carry the battle south. So goes the legend, and indeed, Moorish historians do refer to an obscure battle at a cave in the north.

The earliest historical mention of any religious establishment at the cave is to a collegiate church of Canons Regular of Saint Augustine, mentioned in 1383. A diligent search of all earlier literature reveals no trace of the existence of the shrine, today one of the most famous in all of Spain.[22] The first mention in the Obeso registers is in 1656 and the last in 1803. The apogee of the devotion from the Rionansa valley seems to have been from 1743 until 1783. A disastrous fire, in 1777, put the shrine more or less out of commission for a hundred years, when it was was restored and devotion from the Nansa valley resumed. Persons from the lower valley, which is closer to the shrine, recall their parents making penitential trips to Covadonga on foot, a journey of 80-90 kilometers. Today the personal devotion is confined to older persons of the lower valley. Young people and parish groups throughout the valley make excursions in buses to Covadonga of a patriotic, religious, and touristic nature, but they do not go for promises or penance.

At present the shrine, which is one of the four basilicas in Spain, is a collegiate church run by canons appointed by the Bishop of Oviedo. Our Lady of Covadonga is the patroness of Asturias (the province of Oviedo). Covadonga is a dramatic place—a cave in the face of a steep valley not unlike the Nansa valley.

The Dominican monastery of Las Caldas also attracted valley devotion for its titular image, Our Lady of Las Caldas, during the latter half of the eighteenth century, though much less than either Covadonga or El Brezo (discussed below). Las Caldas is located in the Besaya valley, about 47 kilometers to the east of Obeso. The monastery was founded in 1605. The image, which dates from the thirteenth century, had languished for centuries in a small chapel until in 1683 the Dominicans obtained possession of it for their patroness. Under an energetic prior the monastery sent preachers out to the villages teaching the gospel and spreading, through the establishment of brotherhoods, devotion to the Rosary. At first the monastery depended on the Dominican establishment at Santillana, but soon it became independent. It survived, like the other monasteries in the region, on a combination of paid devotional services for the surrounding countryside, the patronage of wealthy benefactors, and the entailment of property through wills. The image of the shrine became noted for a series of miraculous rescues of drivers and pack animals, for the monastery is located near an old ford in the river, below a dangerous defile. A student of Spanish shrines, Juan de Vilafañe, noted in 1724, "It appears that Divine Providence placed this holy image near these dangerous places so that by Her intercession the travelers are freed from peril, it having been observed that no one has suffered misfortune of any consequence while passing there because her Majesty did not desire it."[23]

The first mention in the testaments of Obeso comes in 1667, the last in 1807. Although the monastery has been reestablished and is now flourishing, its territory is more limited than it was in the eighteenth century, and its zone of devotion now encompasses only the Besaya valley. Many persons go there for confession; a substantial number of marriages, first communions, and blessings of babies are performed there; and about 5000 persons attend each of its two major fiestas—the first and third Sundays in October. But very few of the present inhabitants of the Nansa valley have ever been there. As in the case of La Velilla, St. Toribio, and Covadonga, Las Caldas has not been able to recover in the twentieth century the influence it had at its apogee. The decline of these monasteries was followed by growth of generalized devotions in the homes, and when the monasteries were reactivated, much of their clientele was lost.

IV. El Brezo and La Bien Aparecida

One regional shrine that was popular under the Old Regime has maintained its popularity in the Nansa valley—the shrine of El Brezo, still the most important regional shrine for the villagers. It is located in the mountains of the province of Palencia, about 90 kilometers northwest of Obeso. On the border of the Cantabrians with the Castillan plains, it is as if the valley villagers venture to the very edge of their kind of world when they go to the shrine.[24]

The legend of El Brezo (which means heather) is no less marvelous than those that have gone before. In 1484 two shepherds of the city of Caceres, far away in Extremadura, are sent by a vision of the Virgin to seek her image and found a shrine in the Sierra del Brezo. Eventually they find the site, and with the aid of another vision, they discover the statue by a spring on a mountaintop. An humble shrine is constructed in a cirque at the foot of the peak, then Benedictines from a nearby monastery take over, building an annex on the site.

The legend seems to underestimate the age of the devotion by at least 200 years, for the statue itself is clearly of the thirteenth century. At any rate, Benedictines did have a monastery at the site, a dependency of San Roman de Entrepéñas and San Zoilo de Carrión, and they remained there until they were expelled in 1835, when the image was taken to the village of Villafria, nearby. There it stayed until the shrine was restored in 1850. From 1850 to the present it has been maintained, and the fiesta has been organized by the parish priest of Villafria and one or more majordomos elected by the local brotherhood of the Virgin. It is a remarkable shrine that emits little propaganda, provides few amenities, yet perhaps because of its dramatic site attracts the devotion of a large zone of northern Spain. Unlike Covadonga or Bien Aparecida, El Brezo has received little financial support from governments or Bishops, and it has not been under the care of a monastery for 135 years.

El Brezo first appears in the Obeso records at the relatively late date of 1750 and was very popular from 1766 to 1799, when, like the other shrines, it began to lose support. Presumably the hiatus lasted at least until 1850, but soon after that devotion began again. As was the case of La Velilla, not far to the west, the devotion at least in part may have been due to contagion from nearby Liébana, for El Brezo was until recently part of the Diocese of León. The priest of El Brezo thinks that the Benedictines of El Brezo and St. Toribio may have been in contact, and that the monks of St. Toribio helped to propagate the devotion.

At present the trip to El Brezo is almost uniquely confined to the fiesta of September 21. On that day the Nansa valley bus line sends a special bus up the valley early in the morning to pick up women who promised to go to the shrine and teenagers going to the fiesta dance. In the bus on the last stretch of road from Villafria to the shrine they pass a number of beggars and cripples collecting alms. All those going in the bus confess at the shrine, receive communion, kiss a medal worn by the Virgin, then wait around for the high mass at 11:00 A.M. While they wait they may visit the small shop displaying photos, banners, and booklets about the shrine, and the booths in front of the sanctuary selling food, drinks, and souvenirs. The high mass takes place in the open air, in a natural amphitheatre formed by the mountainsides behind the shrine. There are regularly 15,000-20,000 persons at the mass.

Now, few persons arrive from distant places on foot, but in the past substantial numbers from the Nansa valley have gone to the shrine over mountain trails on promises. Some have made the trip, which takes two or three days each way, several times. Until the Civil War all kinds of gifts were brought as thank offerings to the shrine. The shrine developed, for instance, its own herd of animals, some of which were auctioned off every year. Wax and cloth thank offerings were still brought to the shrine in the late 1950's, when the priests began to discourage the practice. The traditions of penance and mortification are still strong in this mountain region, and women still arrive in the early morning hours or the evening before the ceremony, barefoot and footsore. There was an especial upsurge of penance and promises to El Brezo right after the Civil War. At that time those whose sons and husbands returned safely came to give thanks in fulfillment of promises made during the war. Others came in need, for the economy was disrupted, many persons lost or displaced, and insecurity was rife. The devotion has continued and increased since then, perhaps with less emphasis upon penance and promises since the economy has improved and a social order has been reestablished.

Indeed, the mood of the fiesta in 1969 reminded me of an American college football game. There were six masses in the morning, and masses continued to be said as long as there were people who wanted to hear them. The temple was packed with people going out or in. Women gathered around confessionals and left bunches of votive candles on the altars, whence they were swiftly removed to the sacristy. As I arrived the chaplain was taking the statue down and putting it on its carrying stand. There were at least 20 priests or religious around—those of nearby villages and those native to the region, who had come to help out in the confessionals and in the saying of masses. All together they escorted the Virgin outside behind the church for the open air mass which climaxed the morning.

The mass itself was conducted almost like a pep rally. The chaplain, young, energetic, stood in the pulpit with a microphone directing the crowd. Before the ceremony began he read letters: from the Bishop, blessing all who attended; from the Governor, saying why he could not come; and from the previous Governor, now in Huesca. Then he held up the baton of honorary mayorship, to be presented after the mass. "It cost eight thousand pesetas, gathered from you peseta by peseta so that no one could say that they alone had donated it. It is a gift from everyone, all children of this mother." Then he held up a mantle, embroidered with filagree, and finally he led the crowd in cheers:

> "Viva Cristo Rey!"
> "VIVA!"
> "Viva la Virgen del Brezo!"
> "VIVA!"
> "Viva la Virgen del Brezo!"
> "VIVA!"

Nine priests in luxurious robes concelebrated the mass: three from the three nearest parishes; three diocesan officials; and three members of religious orders who were native to the district. One, the eldest, was a Trinitarian Friar born in Villafria. Throughout the mass when it was time for responses from the enormous congregation the chaplain at the microphone would say, "All right now, everybody together. I said 'Everybody'!" The sermon was a moving eulogy of the Virgin delivered by a Missionary Bishop from Peru. He broke down and wept when he recalled that as a youth he had come to El Brezo to pray for his vocation. The dominant note of his sermon was the love and comfort of the Virgin for us, that the Virgin is the mother of us all. He said he thought it proper that we bring our cares and preoccupations to her and that the proclamation of the Virgin as the Mother of the Church by the Second Vatican Council would help lead to the conversion of the Protestants. As he concluded, he looked up at the multitude on the mountainside and said, "When the churches are not numerous enough to hold all the people, then the world is the church, and the altar is in the heart." During the mass about 500 communions were distributed, bringing the day's total to about 3000.

Afterwards the Virgin was presented with the new mantle ("It cost 40,000 pesetas") and then with the baton of honorary mayorship of the surrounding district of La Peña by the mayors of the towns involved. At this point the chaplain got everybody to cheer, applaud, and wave white handkerchiefs. Finally the image went on procession around the shrine, stopping for a ragged Salve at the front while people kissed the medal and threw money on the carrying stand. Then the crowd broke to the hillsides and the valley bottom to eat the picnic lunches that they had brought and pick sprigs of heather as souvenirs and sacred tokens. At four or five in the afternoon the buses, cars, and people on foot proceeded down the valley, many of them reconvening at the fiesta ground 3 kilometers below where a dance continued well into the night. The bus for the Nansa valley left for home late in the evening.

The shrine of Our Lady of La Bien Aparecida is the most recent addition to the valley's repertoire of more distant shrines. Only in the past 20 years, as far as I can tell, has it been visited by any appreciable number of valley residents. Special circumstances have been necessary to bring it to their attention, for it is quite distant, 115 kilometers to the east.

The shrine started when two shepherd children discovered a small statue of the Virgin, supposedly in the year 1605. (The earliest documentary evidence of this event comes distressingly late—in 1738.) They found the statue in a small embrasure of a chapel devoted to Saint Mark outside the village Hoz de Marrón. People from the village came up to check, and sure enough, confirmed that the statue was there. When they tried to remove it from the shrine it started raining very hard, and they took this as an omen that the image wanted to stay where it was. Later, a neighboring village, missing an image from one of *its* chapels, sued for recovery. The dramatic trial was marked, reportedly, by the paralysis of one of the witnesses trying to regain the statue and was ended when it was discovered that the lost statue had merely been taken to Madrid for restoration. After this account of 1738, later versions of the story transformed the finding of the statue into an "apparition" of the Virgin, and more miraculous elements were added.

By 1906 when a booklet on shrines in the province was published, it was ranked as one of the two most important shrines of the eastern half of the province.[25] At that time, on the occasion of the fiftieth anniversary of the proclamation of the dogma of the Immaculate Conception, a group of priests petitioned the Bishop that the Virgin of Bien Aparecida be named the patroness of the Diocese. The Bishop wisely deferred this decision to the Diocese as a whole. A vote was taken of the archpriests and the representatives in the Cortes, and they voted favorably. It was by no means a unanimous decision, however, because, as we have seen, other shrines were equally important in other corners of the Diocese. But Bien Aparecida, although virtually unknown in the west, filled the need for a symbol and a patron for the region. Santander, as we mentioned above, was a late bloomer among regions. Correspondingly, its symbol came late, long after the establishment of Our Lady of Begoña, in nearby Bilbao (for Vizcaya); Our Lady of Aránzazu, in Oñate (for Guipúzcoa), and Our Lady of Covadonga (for Asturias).

In 1908 the Bishop assigned the shrine to the care of Trinitarian friars, its new status requiring full-time keepers beyond the normal *santero* or *ermitaño*, as the secular tenders of smaller shrines are called. Fifty years later, in 1954, on the one hundredth anniversary of the proclamation of the Immaculate Conception and the two hundredth anniversary of the founding of the Diocese, a movement started to have the image canonically crowned. This formal ceremony, often performed by the Papal Nuncio, must be approved by the Vatican after evidence is submitted to show that the image enjoys the devotion of extensive areas of the surrounding countryside and that the devotion is of long standing. The necessary approvals and arrangements were completed in 1955, which turned out to be the three hundred and fiftieth anniversary of the apparition, and the image was crowned in Santander witnessed by delegations from every archpriestric in the Diocese. These delegations had to bring with them one of their own favored images, as if in so doing to testify to the subordination and fealty of their prized images to La Bien Aparecida.

At this time the conscious effort on the part of the Bishopric to cultivate a sense of identity as a diocese, using the shrine as a vehicle, became more clear than ever. First of all the coronation was proclaimed from every pulpit, and ample information about it reached every corner of the bishopric. Second in a meeting of the archpriests of the diocese shortly after the coronation the following practices were agreed upon:

1. Each parish shall place in an honored setting an image or picture of Our Lady of La Bien Aparecida. (Few now have them, as competition for space on the church walls and altars is fairly intense, and many churches never put them up in the first place.)

2. In the three Ave Marias that precede the litanies in the Holy Rosary there shall always be added, "Our Lady of Bien Aparecida, Queen and Mother of La Montaña, pray for us." (This was done in many of the services I attended.)

3. The Holy See will be asked that the liturgical feast coincide with the popular fiesta commemorating the apparition.

4. In each parish will be initiated the Brotherhood of La Bien Aparecida, all confederated under the Archbrotherhood of the shrine. (Popular devotion in most areas was not sufficient to warrant the enforcement of this provision.)

5. Each year the parishes, on the days that they find most convenient, will make their respective visits to the shrine, and every five years there will be an official pilgrimage of the archpriestric, with the collaboration offered by the Trinitarian Fathers, Chaplains of the Virgin.

6. It is suggested that the day of its fiesta or on another convenient Sunday the consecration to Our Lady be renewed with the words read by His Excellency the Civil Governor on the day after the Coronation.[26]

The younger priests of the parishes of the Nansa have managed to shepherd their flocks to Bien Aparecida, if not every year, at least once or twice so far. And groups from the archpriestrics do make the trip every five years. Furthermore, missionaries preaching revivals in the villages have encouraged the devotion. In these ways the shrine has gained some favor in the valley. There are even cases of promises to the shrine on the part of some of the more virtuosic devotees.

But the shrine stands in contrast to the previous regional shrines described. While all of them, due inevitably to the presence of religious orders, have had an element of propaganda involved in the spread of their respective devotions, this seems to have been exclusively the case for Bien Aparecida in its influence on the valley. As such it is a modern day equivalent of the venerable regional shrines of the rest of Spain—Montserrat, Guadalupe, Covadonga, all somewhat artificially elevated to preeminence. Its accessibility to the valley comes only with chartered bus, and the bus excursions are the only times it is visited. Because of its distance and relative lack of tradition, the shrine cannot compete with more well-known and closer images when it comes to the regular devotions of the valley.

The distant shrines—those beyond the valley—are major establishments. Most are fairly spacious chapels: two or three times the size of the local shrines. They have to be so big to serve their large constituency. The special image that is the *raison d'être* of the church is always located in a central position in the reredos (the bank of images above the altar). For those who wish to pay a special visit to the statue or relic, many (St. Toribio, La Velilla, Bien Aparecida) have a *camarín*, a small chapel behind the altar, and the Virgin can be switched around to reign over this little room. As virtually all of these shrines are or have been administered by religious orders, they generally have outbuildings with cloisters, dormitories, and servants' quarters. As at El Brezo, souvenirs in the form of postcards, medals, or booklets are generally available.

At many shrines at one time or another there has been a house for pilgrims (called a *hospedería* or *casa de novenas*). Here the devotee who had pledged a pilgrimage of nine days, following a pious tradition, stayed for her visit. While the tradition of the nine-day novena pilgrimage has fallen off in most parts of Spain, the custom of coming to spend a spell of time during the summer at a shrine is still common. Often it is accompanied by a spiritual retreat directed by the religious who keep the shrine. St. Toribio, Las Caldas, Bien Aparecida, and Covadonga are all able to accommodate a certain number of pilgrims at the shrine. At Covadonga there are numerous tourist houses and a grand hotel for this purpose. But the persons going to such retreats are generally city dwellers, rarely if ever from the valley we are studying.

The regional shrines frequented by the Nansa valley, aside from sharing the general ecology of the valley, do not clearly fall within one cultural zone, nor do they have well-defined territories of grace. This, I think, is a product of the relative lack of identity of the entire central Cantabrian region. There are little valleys clearly marked off, sometimes big ones. The culture changes abruptly on the southern slopes of the mountains, where dry farming replaces cattle herding. But on the northern slopes the culture changes only gradually as the mountains slope off to the sea. And on the east-west axis, cultural differences, while they may vary capriciously in little ways from vale to vale and valley to valley, show no great breaks at any point. From the western edge of the Basque country to Galicia there is no very great variance, whether ecologically or ethnically. Hence with no clear regional identities to coalesce around, the regional shrines have operated virtually on a market basis. The monks that kept the shrines did not limit themselves to a culturally defined territory in their efforts at propagandization.

The quality of the identity at the different geographical levels marks the quality of the devotion to the appropriate divine figures. We have discussed the relative lack of a sense of membership in a region on the part of the inhabitants of the valley. The quality of identity that does exist at a regional level is that which stems from a shared economy, that of herding. This is not a political or ethnic identity, but rather an identity of habit and artifact that comes through employment in a similar (but not common, as at the village level) enterprise. Hence these regional shrines are not symbols for political or ethnic communities nor are their ceremonies occasions for reaffirmations of solidarity. A better argument could be made that their locations—beside rivers or springs, in caves, or on mountains—are home sites of the forces of nature to which those engaged in herding might logically turn for the success of their enterprises. These regional shrine ceremonies are all at the end of the summer, when people who have had a good summer have the time to take a trip to the shrine to fulfill promises made over the year. But they go as individuals, or as mothers, or as herdsmen, not as members of a region that owes its devotion to the shrine. Only recently have two shrines adopted a self-consciously regional stance: that of Covadonga in the nineteenth century to apply to Asturias, and that of Bien Aparecida in the twentieth century to apply to Santander. These developments would seem to coincide with a cultural reorganization following the establishment of dioceses in a new province.

V. Valley Shrines

Within the Nansa valley, among the villages, an analogous situation holds. The valley is clearly separated from other valleys by mountains, but the demarcations between vales or groups of villages up and down the valley are not so clear. Hence only in those vales or sets of villages that are most clumped and demarcated are there shrines patronized particularly and exclusively by sets of villages.

Four shrines in other parts of the valley have attracted the devotion of the Obeso parishioners since 1600: Our Lady of La Peña, in Celis; The Christ of Bielba; Our Lady of the Luz, of Peña Sagra; and Our Lady of Mt. Carmel, of Cosío (see Fig. 6). That of La Peña, a sister shrine to the Obeso shrine of El Llano, was located just over the mountain of Hoz Alba from Obeso. It seems to have reached the height of its popularity from 1660 to 1675, dropping off

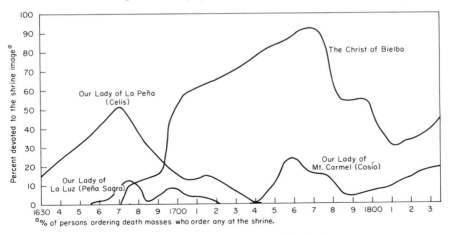

Fig. 6. Graph of Obeso devotion to valley shrines.

thereafter, with a final mention in 1774. One of the three images of the Virgin that are said to have appeared simultaneously on Hoz Alba at the intersection of the territories of Obeso, Celis, and Quintanilla (Lamasón), the shrine of La Peña has now entirely disappeared. When the chapel deteriorated, the image was taken to the parish church, where it was destroyed during the Civil War.

The decline of Obeso devotion to La Peña coincides perfectly with the rise of devotion to the Christ of Bielba. Bielba is the next village down the valley from Celis, and a shrine to Christ of the Remedies was set up there sometime in the sixteenth or seventeenth century.[27] The first notice of this devotion comes in the form of a papal bull, dated 1671, granting plenary indulgences to those who join the shrine congregation. The shrine certainly antedates the bull, but perhaps not as a center of devotion with great popularity, because the first mention in the Obeso records does not come until 1679. Devotion was intense for the entire eighteenth century, and during the years 1741-1778 virtually all persons from Obeso who noted any masses in their wills included Bielba. Like all shrines it declined in the nineteenth century, but it was revived with the aid of Passionist Fathers in 1916 when the ancient bull was discovered, the congregation was reestablished, and a rulebook was issued.

The small devotional was published in Santander and includes an historical sketch of the devotion, a transcription of the bull of Clement X, the text of a novena to the image of the crucified Christ at the shrine, and a hodgepodge of other devotions, accompanied by notations of the indulgences that each devotion earns. (Via-Crucis, Rosary of the Five Wounds of Jesus, Prayers for a Good Death, Sundays of Saint Joseph).

In its historical sketch the booklet mentions "the tradition of the French," the closest thing to a miracle the shrine has had to offer.

> When the irreverent and iniquitous French during their invasion of Spain arrived at the chapel of the Christ, they sought to burn the sacred image. The soldier charged with removing the nails from the crucifix, when he took the one from the left hand, made the arm move, and it hit him hard enough to make him lose his balance and fall on the altar steps, killing himself, and other soldiers who were aiding him in his sacrilegious act.[28]

The chapel where this dramatic event occurred is about 200 meters west of the village of Bielba on a prominence that overlooks the Nansa river. Steep steps lead from the road along the valley bottom up to the shrine, and penitents still go up the steps on their knees in the early morning hours of the fiesta day. That day is September 14, the liturgical feast of the Exaltation of the Holy Cross, the same as that of the Lignum Crucis at St. Toribio. Masses are said throughout the morning, and devotees purchase special ribbons, which boys lift on forked sticks to touch the five wounds of Christ. The ribbons are kept as talismans, worn round the wrist, or hung in automobiles.

Active devotion to the image now extends from Puentenansa in the south to the coast, and from Quintanilla, Lamasón, in the west to Lamadrid in the east. At the high mass, at 12:30, three neighboring priests preside, and the local church choir sings. In the high mass of 1969, about 50 persons received communion, largely women over 60 or under 15. After the service all present, men before women, filed past to kiss the cross. In the evening a large dance was held that drew youth from a zone considerably larger than that of the devotion to the image.

According to many persons, the devotion of the valley to the Christ of Bielba has dropped off somewhat in the past 30 years. The zone of devotion has shrunken, and the intensity of devotion has declined. No longer do persons usually attend from Obeso, for instance.

The third shrine that the Obeso people visited within the valley was Our Lady of La Luz (the Light), a small chapel located on the Liébana side of Peña Sagra. Technically just beyond the edge of the Nansa valley, it is treated here as though it were a valley shrine because of its proximity, and because it shares many features with other valley shrines. As will be seen below, its location and the legend of its origin relate it closely with the more local chapels of the Nansa valley. This shrine is a day's walk from Obeso. It is open only on fiesta days and manned, like Bielba, by the local parish priest. It dates from at least the fifteenth century and has maintained a strong attraction ever since on the countryside immediately surrounding it. The first mention in the Obeso records is 1656, and it exercised a weak attraction, relatively speaking, into the eighteenth century. Obeso, with the other villages of the vale of Rionansa, shares the pastureland of the northern slopes of Peña Sagra (Sacred Peak, literally) and hence has a natural interest in La Luz, which from its start seems to have been a shepherds' shrine. La Luz has been more popular, however, with the Rionansa villages slightly closer to the north: Rozadío, San Sebastian, and Cosío. Persons also go to La Luz from the western villages of Polaciones, and from the vale of Tudanca.[29]

Finally, the Obeso villagers have shown devotion to the most popular image in the neighboring town of Cosío, Our Lady of Mt. Carmel (Nuestra Señora del Carmen). This devotion first appeared in the death records in 1719 and has been moderately popular from 1753 to the present day. Our Lady of Mt. Carmel is the copatroness of Cosío (along with Saint Michael) and her image is located not in a separate chapel, but in the parish church. As Cosío is but a short walk from Obeso and as there is occasional intermarriage between the two villages, it is natural that the devotions of one should spill over to the other. This did not occur until the eighteenth century probably because Cosío did not have the image until then. The devotion was not made an official feast of the Church until 1724. Devotion to Mt. Carmel spread quickly through the northern valleys of Spain in the early eighteenth century, whether due to the efforts of Carmelite Friars, the popularity of the recently canonized St. Teresa, or the result of contact through winter jobs with Cadiz and Jerez, where devotion to Our Lady of Mt. Carmel was particularly intense. The latter explanation seems particularly likely. In the village of Bárcena Mayor (in the Saja valley, to the east) there is a ceramic mural on a house wall portraying Our Lady of Mt. Carmel from Jerez de la Frontera. It was in 1724 that the Brotherhood of Our Lady of Mt. Carmel was founded in the vale of Cabuérniga, also in the Saja valley, in the chapel of St. Anne. Indeed, virtually every vale in western Santander gained an image of Our Lady of Mt. Carmel at this time. The devotion to Mt. Carmel is still extremely strong today. A reflection of this strength can be seen in the fact that when four girls in the village of San Sebastian saw the Virgin in the early 1960's, she appeared in the form of Our Lady of Mt. Carmel.

This inventory of regional and more local shrines was arranged in order of their historical popularity with the parishioners of Obeso partly because the phenomenon of succession thereby became more obvious. The graphs of the Obeso devotions are fairly clear in this respect. When the shrines are arranged by distance away from town, it is seen that as one shrine falls in popularity, another rises. In other words, the shrines themselves do not seem to stimulate devotion or religiosity, rather they are providing a focus or an object for an already extant devotion; or perhaps they are providing a needed service, as a gas station or a drugstore might. Gas stations come and go, depending on the fortunes of their owners, but there always tends to be one around—not necessarily in the same place, but within the same range. The "hidden hand" of the market takes care of that. In the next section, when the philosophy of promises is examined, it will be seen why shrines not only have to be within a certain range, but also have to be beyond a certain range in order to fulfill their usefulness as destinations for penitential pilgrimages.

Why a particular shrine declines and another gains favor is more problematic, and the answers are certainly more complex. In a future study of the life cycle of regional shrines in all of Spain we will search for the answers in more detail, but a few suggestions may be in order here, based on the cases in hand. Novelty and well-publicized claims of supernatural favor—whether in the origin of the image or in the miracles it performs—must play their part in the rise of shrines. The apportioning of indulgences by the Church also was an important factor in certain epochs; in the case of Bielba the papal bull seems to have been decisive. Shifts in regional identity make some shrines more attractive. But these factors do not explain the decline of the shrines replaced by the rising shrines. While in some cases, like the fire of 1777 at Covadonga, such declines may have been haphazard, there may be some larger principles at work. It has been suggested that new saints replace others that have the same function when the first saints are seen as useless, as with a succession of plague saints that follow one another after major epidemics. If shrines are predicated on divine help and divine help is never forthcoming as much as might be wished, would this not occasion a fairly regular turnover of shrines? Under this dynamic, shrines would be "worn out," as it were.[30]

Other factors are at work. It will be noted from reading the preceding inventory, studying the graphs, and perusing shrine histories that on the whole shrines experience, throughout Spain, periods of expansion and contraction. The last half of the seventeenth century through the beginning of the eighteenth century, the beginning of the twentieth century, and the postwar period have seen the expansion of shrines. Periods of decline—in the beginning of the nineteenth century, and in the 1920's and 1930's—correspond not only to liberal governments, but also to periods of economic crisis, when there was simply not enough money for the people of Obeso, for instance, to will offerings to shrines. These cycles, and their national implications, will be dealt with at length elsewhere; it suffices to note here the role of economic hardship in cutting off the income of the shrine. Economic crises might lead to changeovers of shrine devotions. In the hiatus, the tradition of worship at one shrine might weaken sufficiently for a new shrine to step in when the economy loosened up. It is possible that this happened for the valley in 1700, when a new set of devotions seem to have taken over.

Note also that the periods of contraction in shrine devotions—the temporary retreats from regional shrines, as it were—correspond to those contractions of work excursions outside the valley. While both contractions are probably related to a depressed economy, they may also be related to each other. Work away or trips away may be related in a number of ways to devotions to shrines. Promises were made to go to shrines by women on behalf of men fighting in the Civil War; promises may also have been made regularly for men and boys going far away from home to work. We have also seen how in certain cases men working away brought new devotions back to the valley: Mt. Carmel from Andalusia and Pilar from Zaragoza. Although I have little evidence that would support this factor, it may also hold in regard to the regional shrines—they may have been devotions men picked up or fostered on work trips in the shrines' vicinities.

A final kind of contraction and expansion has to do with the shift in identities discussed in the first chapter. In the last 15 years the more regional shrines like El Brezo, Bien Aparecida, and Covadonga have grown, while the more local shrines, like Bielba and the village shrines discussed below have, if anything, contracted. This may be a consequence of the increasing cosmopolitanization of the countryside. The more touristic shrines—the shrines that capitalize on provincial, diocesan, or cultural identities— are flourishing throughout Spain as never before. The more local shrines, unable to provide amenities for buses and cars, often inaccessible except by foot, and having as their territories of grace small units that have been superseded and rendered anachronistic as foci for identity (the vale and to some extent the village), cannot compete. Their magic leaks out. Spain is rapidly changing from a patriarchal rural society to a society that is geared to the pleasures and tastes of an urban middle class. The Church, too, has reoriented its priorities and its facilities to service and cater to this group. Small rural shrines are no longer chic, now less attended to than they were directly after the war, when they epitomized the values of the new regime. During the years of Republican strength in Santander many of the churches were desecrated, their images burnt. The story is told throughout the valley how a militiaman from Liébana went to the shrine of Our Lady of La Luz, broke into it, and took out the statue to shoot it. The bullet ricocheted off the statue and cut off his testicles. This parable corresponds to that of Bielba and the French soldiers. Immediately after the war the youth of San Sebastian, with their priest, had a pilgrimage to La Luz in celebration of the victory of the Nationalists. But as the passions subsided, the value of the local shrine as a symbol for a political position also declined. And as the nation in the late 1950's began to revive, the expansion was chiefly centered on the cities. A new cosmopolitan ethos took over that made the local shrines seem less important.

An anthropologist reports a similar finding for Buddhist Ceylon: "Better roads and better communications have made the pilgrimage so easy that the grand round is now within the reach of most villages. With the decay of caste and feudalism and the democratization of the state, there is, I think, a tendency toward the universalization of the pantheon, e.g., the village and regional deities are losing their authority and the guardian deities are taking over their powers."[31]

But although the village shrine is at present losing out, it has been in the past, and for some persons continues to be, the chief focus of their devotion. As a coordinated working operation, a focus of social activity, a parish, and a unit of government, the village has been a major source of identity. In fact, family names are often derived from the names of villages in this region. The shrine image of the village epitomizes the identity as a villager and is usually considered to protect the enterprises of the village community as well as the villagers separately. The descriptions of the village shrines in the valley and their ceremonies should convey some sense of what shrines mean to people who live near them.

The devotions of the Obeso people to other shrines have always been in addition to their devotion to the shrines within the village territory. Such devotions were enforced by the parish priest, who made it his business to ensure that for every vecino who died, the inheritors or members of the family would have a certain number of masses said (and paid for) at each of the shrines within the village. The most important of these in Obeso was (and is) the chapel of Our Lady of the Lowlands (El Llano). This shrine, situated 500 meters to the south of the village on a lower portion of the valley slope, existed long before the death registers began to be kept. An image dating from the thirteenth or fourteenth century was destroyed during the war. The chapel is narrow and bare, about one-half the size of the parish church. It is normally locked, like most local chapels, only open when the priest is there saying mass.

Its legend and its location are discussed below with the other valley shrines which, with La Luz, seem to share certain important characteristics. The devotion of the villagers of Obeso, Cosío, and Puentenansa to El Llano seems to have been of long standing. Like many chapels in the north of Spain, this one has been rebuilt several times with subscriptions from villagers living in the New World.

The Llano shrine ceremony on August 15 includes a mass celebrated by three neighboring priests, a sermon, a special choir, and folkdances *(picayos)* performed outside in front of the statue of the Virgin. There is no picnic since the shrine is reasonably close to three villages and families can go to homes for lunch. On other days during the year any family may commission a mass to be said at the shrine, and on many evenings women say the rosary at the shrine door.

As with all other shrines described, Llano was maintained over the years by a combination of offerings and deeded lands. In 1691, according to the shrine records, the Virgin possessed small parcels of land in Cabuérniga, Lamasón, and Obeso to the value of 160 ducats. The income of that year came from gifts of maize (12%), offerings of money (10%), and income from land rents (78%). From this income (and leftover capital) a certain number of masses had to be paid to the priest, wax had to be purchased, and small fees were paid for the collection of rents, for the keeping of accounts, and to the notary who transcribed the Episcopal visit. At that time masses were said at the shrine every Saturday. The Episcopal visit of 1709 noted the presence of a "picture of a miracle," clearly an ex-voto painting, and ordered it removed, evidence that the struggle against folk offerings and in favor of the dignity of the sanctuary is an old and constant one. The income of Llano from lands (as in all Spanish shrines) was cut off in the years 1835-1870 as the government sold entailed property. The practice of deeding land to the shrine has not resumed in recent years.[32]

Just as Obeso had Llano for a local shrine, so virtually every village in the valley had a shrine image, almost always that of the Virgin. But none of them, with the exception of Bielba and La Luz, attracted devotion from any farther than the adjacent villages, and so they show up rarely in the Obeso death records. Those that do, occur in the testaments of persons who had moved to Obeso from the village of the shrine in question or who had relatives there.

That every village should have a shrine of sorts is not surprising, because the village has been the major kind of identity in the valley. While the region and the vale are somewhat amorphous and ill defined as identities, the village is a very clearly, perfectly bounded concept, one that is particularly apt as the locus for a divine patronage.

Like El Llano, some of these sacred images are in isolated chapels. Like Mt. Carmel in Cosío, some of them are venerated in the parish church. Those in the parish churches, with one exception, seem to be of more recent origin (seventeenth and eighteenth centuries). Such shrine images, whether in a chapel or a parish church, have a clear ascendancy over the other images in the parish. Their fiesta is celebrated at least by a High Mass and usually by a set of holidays. Many personal problems within the parish, and grave communal problems, are offered to the special image for help.

Village parish	Active patron	Shrine type
Polaciones	N.S. de la Luz	Chapel
Puente-pumar	N.S. de la Puente	Parish church
La Lastra	N.S. del Carmen	Parish annex
Tudanca	N.S. del Vado de la Reina (=N.S. de las Nieves)	Chapel, parish church
Santotís	N.S. de la Vega	Parish annex
Sarceda	(Las Animas)	Parish church
San Sebastian	(Las Animas)	Parish church
Rozadío	(Las Animas)	Parish church
Cosío	N.S. del Carmen	Parish church
Obeso	N.S. del Llano	Chapel
Puentenansa	N.S. del Salud	Parish church
Carmona	N.S. de las Lindes	Chapel
Celis	N.S. del Carmen	Chapel
Bielba	El Sto. Cristo de los Remedios	Chapel

Except in the cases of San Sebastian, Sarceda, and Rozadío, in which devotion to the Souls in Purgatory seems to take the place of an active patron, and Bielba, where the Christ is clearly dominant, all the other villages have as their active patron the Virgin in one or another of her advocations. The active patron, as we have defined it, is the village shrine image. It should be distinguished from the formal patron, the titular image that gives the parish its name. The titular image, usually a Saint (Michael, Peter, Facundo, John, Blaise, Ann, George, Roch), receives little or no devotion now. The only exception is one titular image of the Virgin that is also the active patron, that of Puente-pumar. Few if any masses, whether supplicatory or in thanks, are said to the titular images, and as far back as the records go, that is, as far back as 1600, this seems to have been the case. The only kind of exception would be for certain specific ailments or conditions that the Saint was said to prevent or cure. In this respect they were treated just like the other saints in the church and in the subchapels of the barriada.

At this point we must abandon the death registers of Obeso, for they do not enlighten us about the local shrines in other villages in the valley. One of the most important of these is the shrine of Tudanca. As in the case of El Brezo, La Bien Aparecida, and the Christ of Bielba, a closer look at this village shrine and its annual ceremonies may help convey a feeling for the relation of an image to its territory of grace. The shrine is referred to both as Nuestra Señora de las Nieves (Our Lady of the Snows) and Nuestra Señora del Vado de la Reina (Our Lady of the Queen's Ford). The former refers to its feast day, which is August 5, the day of a famous shrine in Rome. The latter refers to the shrine's location at the crossing of a stream. The chapel is in the mountains 5 kilometers behind the village of Tudanca on an old post road that once linked Cabuérniga with Polaciones.

Like other shrines in the valley, this shrine has had spells of popularity and spells of abandonment. At present it is somewhere between the two extremes. Vado de la Reina shows up rarely in the Obeso records (although as early as 1626), and Llano, in turn, has rarely been visited by the people of Tudanca. Both are shrines that have attracted, at most, neighboring villages. Both are basically village images acting, in their most popular moments, as the protectresses of the cluster of villages that comprise the vale. In their least popular times they attract little more devotion than that of the home village.

In the nineteenth century an annual cattle fair was held around the chapel on the fiesta day. Although the fair died out, the fiesta remained a popular one into this century. It lapsed under the inattention of one disinterested priest, then became popular again immediately after the war when the company building the nearby dam offered to rebuild the chapel. Now on August 5 a mass is said in the chapel, but there are no other festivities.

The image remains in the remote chapel every year until the first week of May, when it is brought down to the parish church. There it stays for three months, an object of great devotion, surrounded every Sunday with many candles, until July 25, the fiesta of Santiago. The image's stay in the parish church permits the old people in the village and others who are too inactive or ill to climb to the chapel to visit the Virgin. Since it comes down in time for the plowing and sowing of the mies, perhaps its descent was once related to the Virgin's role of protectress of the village fortunes because the mies in earlier times, as it provided much of the year's flour or corn meal, was critical for the village survival. Two other chapel images in the valley, Llano and Vega, were placed in or overlooking the mies. But this link is not made explicitly by the villagers.[33]

On July 25 at 4 P.M. the church bell rings for the second time that day and the ceremony of leave-taking for the image's return to its shrine occurs. I witnessed this ceremony in 1968 and 1969. Prior to the beginning of the rosary, the older women in the village knelt one by one in front of the statue, and the girls kissed the ribbons hanging from the old wedding bouquets that had been left in the Virgin's hand. The men gathered outside the church and the women sat inside with the children. After the priest left the men to go inside and begin the rosary, children came running out, followed by the statue carried by women; in 1969 it was carried by the son of a woman recently deceased. They continued carrying the statue up through the village, preceded by the men with bared heads, teenagers firing rockets, and the ringing of the church bells, the bell at the casona chapel, and the bell in the parish annex of La Lastra across the valley. Two women sang the responses to the rosary, and the procession advanced in slow stages, stopping for each mystery.

In a procession as in Tudanca on July 25, all of the village takes part. Hence it is not a parade put on for the benefit of spectators, but a kind of pageant for the benefit of the participants themselves, and above all for the image being carried in the procession. The order of these processions dates from at least the seventeenth century and probably long before. Children go first, then men, then the priests and the image, and last the women. In medieval cities the processions were divided into guilds. In the villages, where the people all do more or less the same work, the corporate divisions are by age and sex. In both cases the procession forms an image of the way the social unit views itself, or is encouraged to view itself, as an organic whole made up of distinct parts. The statue itself is a part of the organic whole, a member of the village. As these processions are held only for the acknowledged patrons they are at once ceremonies in which (through, above all, the picayos songs) the people proclaim their fealty to the patron and their gratitude for its protection and are at the same time demonstrations of the extent to which the patron is an active village member.

As images of social wholeness, the processions have an added significance. The villagers for once in the year see the village as a social unit, abstracted from the buildings and the location that make it a geographical unit. "As in the church itself the spectators were also communicants and participants: they engaged in the spectacle, watching it from within, not just from without: or rather, feeling it from within, acting in unison, not dismembered beings, reduced to a single specialized role."[34]

At a point about a third of the way up the mountain that overlooked the village and the other villages in the valley, the priest read the sixth mystery and everyone sang the Salve Regina. Then the men sat down for a rest and a smoke before going back down to the village, while the women, some of them, went on with the statue to its chapel in the mountains. The older ladies had not gotten very far up the mountain with the procession, nor had the older men. I sat with them afterwards at the bottom of the hill in the village, and each oldster came by and reported how far he or she had gotten. "If we have a true love for the Virgin and do the best we can, how can it matter how far we go?"

In 1968 I had gone on with the women to the shrine. As the path snaked up to the pass, more and more people dropped out, especially at the turns. At the top of the pass about 50 persons were left, and there was a little ceremony where an oratory used to be. The women who turned back kissed the statue, and the procession, now reduced to a core of about 15, continued intact to the chapel deep in the next valley. One woman had been to a sanatorium and had promised to carry one handle of the traveling stand to the shrine. Her husband, the only man who continued beside me, came along to spell her. The procession was silent except for frequent Ave Marias, Salves, and the singing of the shrine hymn. The rosary was sung with responses in order to give breath, especially on the climb.

At the shrine the middle-aged unmarried woman who was more or less in charge opened the door, and they cleaned out the chapel. First there was a Salve outside the chapel, then another inside. Only when the statue was placed on its altar in the chapel, however, did they kneel to her. The woman in charge went and gathered greens, arranged them with plastic flowers, and did the final placing of the statue. Then the women kissed her shoulder, her hand, the baby Jesus' hand, and the baby's loin. They clearly loved her and were sad to leave her. "Until the first of May, Mother, with God's will."

The return trip was gay and festive. They sang secular songs, pop songs, and well as the hymn of the Virgin. They said it was the same every year, that it has always been the same ones who continue on. (Although I later learned that until about 1940 the entire village, men and women, would make the trip to the chapel.)

In both 1968 and 1969 I attended mass at the chapel on August 5, the feast of Our Lady of the Snows. The image is known by this name in the rest of the valley. It is a devotion common to many of the more mountainous regions of Spain. The first year I walked up with the priest and was surprised to see that we had to wait until the woman in charge of the key arrived before we could get into the shrine. Both years about a third of the village turned up at the shrine for the ceremony. In addition to a cross-section of the village, there were several families of migrants, villagers now living elsewhere.

Before the mass, women clustered around the entrance of the tiny chapel, taking in votive candles and helping set up the service. The "Dons"—the local priest, the emigrants, the valley professionals—stood in a circle near the chapel, while the herdsmen and boys formed their own group farther back. About 15 of the 60 active men in the village were present, half of the active women, all of the teenage girls, and half of the teenage boys. During the mass the older boys spent their time, like me, looking at older people, but all joined in the singing of the hymn to the Virgin. In 1969 the sermon was given by the young liberal priest of Polaciones. In resume, he said:

> The hard climb to get here is much like the Christian life. In the Bible we find time and again the symbol of the road, the mountain, and the climb. Life is like a road, but on the road we get dirty. Anyone involved in life, on a trip, gets dusty, dirty. Coming to the shrine we come to a place where we can become clean, clean as the snow of Our Lady of the Snows. But after this we must go back down the mountain. We must go down below, bringing what we have learned back to the active life.

After the mass everyone sang a Salve, and the priest concluded with the traditional formula, "Hail Mary," the congregation replying, "Conceived without Sin." The priest came out and formed a new conversation circle while the women and the girls squeezed into the chapel for a rosary. In the circle the men recounted famous traffic accidents and remembered notorious drivers (especially an old priest) while the rosary proceeded in the background. Then everything broke up, some of the herdsmen taking advantage of their nearness to the grazing grounds to check on their cattle and horses, and the priests, notables, and women returning to the village.

Back in town most families had a festive dinner, especially those who had relatives visiting for the occasion, and in the evening the men reconvened, as usual, in front of the television set in the bar. In former times a dance would have been held, in the village or at the shrine, and after mass the village youth would have danced the picayos in honor of the Virgin.

There are two other chapels that have much in common with Luz, Llano, and Vado de la Reina. One is close to Tudanca, in the hamlet of Santotís. There the Virgin of the Vega (cultivated land along a river) is venerated. Formerly the image presided over a chapel on the riverbank below the village, where ruins are still in evidence. At present the statue is in the parish annex (Santotís, like La Lastra, is officially in the parish of Tudanca). The fiesta day, now celebrated with little enthusiasm, is August 15, the same day as Llano in Obeso.

The other chapel is that of Our Lady of Las Lindes (boundary lands) of Carmona. Although Carmona is legally a village pertaining to the Vale of Cabuérniga, geographically it is part of the Nansa river valley. It is closer to Puentenansa than to Cabuérniga. Hence it is included in this survey. The chapel of Las Lindes is high on the mountainside above the village, and many of the Carmona villagers are deeply devoted to the image. The chapel mass is on August 18, the day after the fiesta of the copatron, St. Roch. Here, as at Llano, it is the custom to dance the picayos to the Virgin, although in most recent years the custom has been suspended due to a disagreement between the village youth and the priest.

The devotion to the image is handed down from mother to child, like that of Llano and Vado de la Reina. The shrine, it seems, has never attracted devotion from beyond the village, although within the village the devotion to its image is more intense than anywhere else in the valley. It is possible that this intensity is related to the intensity of Carmona's sense of distinctness from its neighbors. It has always been a distinctive village, and there are many colorful stories told about its inhabitants. Here a sense of identity seems to be epitomized and symbolized by an image that stands for that identity. And the degree of unanimity and intensity of the devotion seems to be related to the degree of unanimity and intensity of the identity.

Considering Luz, Llano, Vado de la Reina, Vega, and Lindes together, there seems to be a pattern to the location of these oldest chapels in the valley and the locations of the divine appearances reportedly at their origins (see Fig. 7).

The typical case is that of the images venerated (or, in the case of Celis, formerly venerated) in Obeso, Quintanilla, and Celis, known respectively as N.S. del Llano, Santa Maria, and N. S. de la Peña. These three statues are said to have appeared simultaneously on Hoz Alba, a mountain that stands at the intersection of the boundary lines of the three village territories. The statues are said to have been facing their respective jurisdictions. Three other chapels seem to follow this pattern. That of Lindes, although it has no tradition of apparition, is located only a little below the ridge dividing Carmona from the next village. And its very name draws the relationship with the boundary line. Las Nieves at Tudanca is said, in the shrine hymn, to have appeared on "aquella peña" ("that peak") which, in fact, marks the boundary line with the neighboring grazing preserve of Campóo and Cabuérniga. And La Luz reportedly appeared to a shepherdess on Peña Sagra, a peak which serves as a four-way boundary between the vales of Lamasón, Liébana, Rionansa, and Polaciones. What's more, Peña Sagra formerly served as a boundary between the four bishoprics of Burgos, Palencia, León, and Oviedo. The gravitation of apparitions and shrines towards boundaries seems to be an occasional phenomenon occurring throughout Spain. In local terms there might be several explanations. One, pertaining especially to herding cultures, is that a certain amount of anxiety and interest focuses on the boundary because it is the line over which livestock must not stray. If they are caught over that line, on foreign grazing land, they are liable to being taken to the corral of the next village and held there until their owner comes to pay a fine and the cost of their maintenance in the corral. The legend of La Luz makes this point almost explicitly, for when the Virgin appears at Peña Sagra, she helps the shepherdess, who has lost her sheep in the fog, to gather the sheep, and then the Virgin takes care of the sheep while the shepherdess goes to the village below to report the apparition. In other words, the Virgin keeps the sheep from getting lost and also from going over the top of the mountain. This theme is found in several other shrine legends throughout Spain. The other legends in the valley are not so explicit; most of them are fragmentary, but most include the item that it was a shepherd or shepherdess to whom the Virgin appeared.

Another aspect of the gravitation of apparitions and shrines toward the village boundaries is more universal throughout Spain. The gravitation points to the very essence of the idea of patronage: The shrine belongs to one village; it

Fig. 7. Map of apparitions and chapels in the upper Nansa valley.

epitomizes that one village; and it protects with special favor one village as a collectivity. Thus the apparitions, while they often occur very near the boundary, only with utmost rarity occur on the boundary. Always a little to one side. In fact, it is as if the image marks off the outer edge of the village territory saying, "This is ours," much as a fence post or a stone marks the limits of one man's section of a meadow.

After appearing for their respective villages at their respective locations, the apparitions take different paths. But the legends of Llano, Vado de la Reina, Luz, and Lindes all include the item that the villagers attempt to build the chapel for the statue in one place, and at night two oxen mysteriously transport the materials to another site. In the case of Luz and Lindes the rejected site is lower than the present site; in the case of Llano and Vado de la Reina the attempted site is not mentioned. This legend item is frequent throughout the Cantabrians and the Basque country. It corresponds to a common element of Catalan legends in that there is often a tension between higher and lower places, mountains and rivers. In the Catalan legends the location of the statue is shown by an ox running from river up to mountain, and back.

Two lessons seem implicit in this legend item: first that the shrine has a supernatural rationale for being precisely where it is: It is divinely chosen. By implication, the site as well as the statue has something holy about it, perhaps is worthy of some devotion in and of itself. The second lesson is the attention accorded in the legend to the placement of the shrine up or down the mountainside. The people attempt to build the shrine lower down because it is easier to do so. But by the patent wish of the Virgin, the shrine must be built higher. The people must expend more effort to build it and more effort to attend to it. The shrine will be more useful as an object of penance because it is harder to get to. Thus the sacrificial aspect of devotion is introduced into the legend. A chapel or a patron must be earned to be deserved; grace itself must be purchased with toil. And, most important, people must accede to divine will when that will is made manifest, even if it means inconvenience.

In addition to this relatively homogeneous group of chapels, whose origins are lost in the early history of the valley and which share certain elements of legend and location, the valley has seen the establishment of more modern chapels. The earliest of the new series is that of Bielba, discussed above, which originated in the sixteenth or seventeenth century when the cult of the crucified Saviour spread across Spain.

The next most recent chapel was built in Polaciones in 1822. Polaciones is an isolated vale with a clear identity. It is a wonder, therefore, that it did not have its own shrine, in addition to the various parish churches (of which there are nine), until such a late date. Apparently the image was located in a school, and a wealthy inhabitant put up money for the construction of a separate chapel. The name and the fiesta date of June 24 were taken from the more famous shrine on nearby Peña Sagra. The number of coins tossed through the grill in the locked door attests to a certain devotion on the part of the valley, but on the whole the little shrine has not been very successful in eliciting the vale's attention, this in spite of its location on the main road, or because of it.

Without exception, the chapels in the valley are of a humbler nature than those with regional attraction beyond the valley. Generally the valley chapels consist of one oblong wing, the head facing east and the entrance facing west. Made of stone and mortar (and as such, slightly more elegant than the stone houses of the villages or the stone winter barns) they are generally quite barren inside, aside from the altar. There may be one or two benches, and an occasional prayer chair that belongs to a regular devotee, but the walls are bare. The image of the shrine is on or above the altar and is generally accompanied by one or two other statues of local devotion. The Luz shrine has an ornate reredos, but it is unique in this respect. All of the shrines are small, with a capacity for no more than 200 nor less than 40 persons.

Indeed, they offer quite a contrast with the parish churches. The parish churches are considerably larger, because they must accommodate the entire village population of 200-300 persons every Sunday. They are also considerably more decorated, with many more images, Mission crosses, religious pictures, the stations of the Cross, plus, of course, the full paraphernalia of the mass.

The nature of the shrine devotion when the revered image is located in the parish church rather than the isolated chapel, is, of course, somewhat different. In the chapel the devotion received is episodic. It is limited to the fixed days in the year that masses are said, the other days that people hire the priest to go there and say special masses, and those occasions that lay people stop at the chapel, whether to say a rosary, merely to take off their hat and stand in silence, or perhaps toss a small coin through the grating on the door or window. Such sporadic visits become more frequent as the shrine is more accessible. The revered image in the parish church is the extreme case. Even though some of the parish churches are locked during the day, the key is usually at the nearest house, and the image is easily accessible to those in need of aid or comfort.

Yet the use of a parish church as a site for a revered image has its drawbacks. For one thing, because the parish churches are centrally located, or at least located so that one must go through the village to get to them, the act of devotion can no longer be a private one. In villages where one is looked down on for being too devout (a kind of "rate buster," as it were) as well as too little devout, this is a drawback. Another unfortunate aspect of the lack of privacy is that it contravenes one of Christ's precepts—that one should pray not in public, but away from the view of others. This precept has had its effect in these villages; especially the older women seem to believe that it is better to pray where people cannot know you are praying. For then the temptation to exhibit one's holiness is entirely put aside, and one can be sure of one's motives. Of course, there is literally nowhere, except for the private rooms of one's house, where one can go in this countryside and be unobserved. But the shrine, set apart, offers more chance of freedom from the public eye than the parish church.

In another way the location of the parish church in or near the village inhibits devotion. A chapel, apart from the village, is more public property, and persons from other villages feel less inhibited from visiting it. In this valley each village, as we have remarked, is nearly a tribe unto itself, and there is always a certain degree of tension involved in visiting from one village to another unless one has relatives there. Fights break out among people of different villages. The church and the village image are included in this territorial possessiveness. The effect of this factor can be seen in those villages that have chapels but where the image regularly spends part of the year in the parish church. Devotional visits by outsiders and other visitors to the image are generally limited to those times when the image is in its chapel, not the parish church.

In the case of revered images within the parish churches, or shrines within the village agglomeration, therefore, the devotion of people from other villages seems to be restricted to fiesta days of the image in question. At that time there is a kind of tacit truce, and outsiders can be present. Again, the nature of the devotion to an image in the parish church differs from that to an image in a chapel. Devotion to an image in a chapel, especially if it is located out of the way, can reasonably be considered a penance—in some cases, even on the part of the home villagers—while visiting an image at the parish church is not. Images revered within the parish church are objects of sacrificial devotion only from persons outside the parish, who must exert effort to reach the parish. Within the parish the devotion is likely to be supplicatory, involving novenas, or of gratitude, but not involving promises. Hence, for instance, the typical visit to a chapel by a villager might be one of gratitude, or perhaps an intentional mass, while the typical devotion to an image in the village might be a series of prayers over a number of days or the more personal discursive kind of prayer-consultation of someone with a problem or in anguish.

With the exceptions of the parish churches of Obeso and Celis, recently remodeled in an austere simplicity, all of the parish churches and parish annexes in the valley have a number of images on the altars—anywhere from five to fifteen. Four or five of the churches have in them an image that attracts a special amount of devotion, equivalent, with the reservations noted above, to that received by the shrine images in the chapels. This image differs from the others in the church in that it is attributed special powers, usually rather generalized powers; in other words, it acts as an all-purpose protector, or patron. In addition to that of Mt. Carmel in Cosío, whose fiesta is celebrated on July 16 with a high mass, a solemn procession around the village, and a dance, the image of Salud (Health) at Puentenansa is the only other of these images that draws more than the devotion of the village. Women from several neighboring villages attend mass there on September 8, the feast of the birth of Mary.

As in the case of the chapels, all of these active patrons are forms of the Virgin Mary. But in almost all of these cases the parish church where they abide is not dedicated to Mary, but to a Saint. This may be an indication that devotion to Mary became paramount only after the parishes were founded. If devotion to the titular Saint ever existed, it has been virtually entirely supplanted by devotion to one or another image of Mary, which has, whether officially or de facto, become the copatroness of the village.

It should not be mistakenly supposed that because a village possesses an especially revered or powerful image, each individual in the village reveres that image more than other images. In Puente—pumar, for instance, as many persons venerate La Luz as venerate the image in the parish church of Nuestra Señora de la Puente and many have no special devotion for any image. Likewise in Santotís devotion is quite evenly divided between Vado de la Reina in nearby Tudanca and the local image of Our Lady of the Vega. Cosío, too, seems to be divided between devotees of Mt. Carmel and Llano of Obeso. Tudanca, Obeso, Puentenansa, and especially Carmona are fairly unanimous in devotion to their own active patrons.

The shrine images in the chapels and to a lesser extent the active patrons of the village churches are the objects of devotion of the village as a whole. In times of crisis the villagers have turned as a corporate group to these images, to the divine figures that these images represent, as their special helpers. They have also turned to them on regular occasions, the fiesta days throughout the year, as mediators of the ordinary graces considered necessary for the successful fulfill-ment of the cycle of the year, the agricultural cycle. All of this collective worship, often stimulated by the Church itself, has by no means preempted the usefulness of these sacred figures for families and individuals. Indeed, the shrine images are among the most common choices for individual petitions and protec-tion. But they are supplemented in their aid to families and individuals by the generalized devotions, some of which are to be found in the barriada chapels, others in the parish churches, and still others in the homes.

VI. Barriada Chapels

In addition to the shrine of Our Lady of the Lowlands (Llano), the patroness of the village, another sort of shrine shows up in the Obeso records. This is the small oratory chapel, usually located in houses, or converted stables, almost always within the settlement proper. It is the smallest of the shrines, the least significant of the located images of the valley. In fact, here the distinction between a shrine and a generalized devotion becomes a little blurred, as some of these little chapels could fall into one category and some into the other. But to the extent that they were used as devotional centers for the barriadas around them, they would be classifiable as shrines. In Obeso there were several such shrines, and they seem to correspond to the different barriadas of the village. Although throughout the villages in the valley the inhabitants know where these chapels were, today few of them survive. Most of them, established as pious works by wealthy citizens, have reverted after a time to the family of origin and been reconverted to secular uses. This reconversion seems to have taken place above all in the early years of the twentieth century.

At the time the records began to be kept in Obeso (1615-1630) two of the chapels already existed, one devoted to St. Roch and St. Lucia and another devoted to a shepherd saint, San Mamés, both in the barriada of Rioseco. In 1655 a nobleman died, willing all his property to his wife, "to build a chapel in front of his house that both of them had planned to build, dedicated to St. John and St. Catherine" (their namesakes). This chapel was never built. But in 1680 the parish priest constructed a chapel in his house in the barrio of Pedreo and founded chaplaincies there, that is, he endowed a number of masses to be said there each week. This chapel he dedicated to St. Anthony the Abbot, protector of livestock. The last of these minichapels was set up in the barriada of Obeso proper around 1750 by a nobleman in his house; it was dedicated to Our Lady of the Immaculate Conception. At the time of his death in 1759, the nobleman, who had lived in Andalusia, founded a chaplaincy for one of his nephews, with masses to be said in the Conception chapel two days a week, on a capital of 45,000 reales. At present only the St. Roch chapel still exists, and even it is almost always locked. The others have reverted to being stables or storehouses. These chapels were maintained both by their endowments and by the rule that all persons dying who are vecinos of a parish must pay, as part of their funeral, for a mass at each chapel. Personal religious foundations of this nature were prohibited throughout Spain in 1763.[35]

Like Obeso, the other villages in the valley had their intravillage chapels. San Sebastian had two, dedicated to San Ramón and the Virgen de los Remedios. That dedicated to San Ramón was set up around 1740 through some connection with the Mercedarian order, for it was stocked with Mercedarian Saints. The chapel reverted to use as a stable in this century, and the image of San Ramón was moved to the parish church. It was a common practice for women in the valley who were anticipating a difficult childbirth to send a candle up to the chapel for San Ramón, the patron of difficult childbirth. The chapel of the Virgin of the Remedies still exists—a bare little room with a crude statue on the altar and some plastic flowers. The Corpus procession makes a halt at this chapel, as it used to at the San Ramón chapel.

The other chapels I know about within the villages are as follows. All were founded prior to 1700. They were founded most likely by the more wealthy citizens.

Celis: Our Lady of Mt. Carmel
                                    San Antonio

Cosío:	San Juan Santibañez
	San Miguel
Puentenansa:	San Roque
	San Justo Pastor (later changed
	to Sacred Heart of Jesus)
Tudanca:	Our Lady of Cocharcas
	Our Lady of (the Immaculate) Conception
Lombraña (Polaciones):	San Juan Bautista
San Mamés:	San Miguel
Carmona:	San Antonio Abad
	Cristo

The effect of these chapels was to permeate the villages even more with presence of the holy. They were substations of the parish church in each barriada, dedicated to a saint not usually present in the church. They also had their equivalents in the countryside, where, in addition to the shrines, there were little oratories at crossroads, at passes, and at the entrances to the mies on the roads going into the village. Many of them were undoubtedly erected in gratitude for graces received by wealthy individuals from the image in question. In the case of the chapel of Cocharcas, in Tudanca, the chapel was built to house an image brought from Peru by a returning Indiano who had been saved by the statue from a landslide. While those within the village were generally built as private chapels, they came over time to acquire the allegiance of the barriadas in which they were placed.

VII. Images in the Churches

Until perhaps 20 years ago, the home was a major center of worship, and most families maintained regular devotions in the evening. It was the custom throughout the valley, dating from at least the eighteenth century, for the father to lead the saying of the rosary around the dinner table before or after supper. This was done in spite of the likelihood that the women in the home had already attended the village rosary service. The rosaries said in the home generally included a number of special mentions of saints or devotions particularly favored in the household. Now there are very few families in the valley that recite the rosary together—perhaps only one or two in each village. Nevertheless other family devotions continue. Mothers generally recite bedtime prayers with their children until their children are about 11 or 12. Each household abounds in religious pictures, little plastic statues, and religious calendars. In addition, most households subscribe to at least one religious magazine, often the only printed material that regularly comes into the house. For most families cultural material was mediated largely by the Church until radio and television started bringing a more secular outlook.

Just as the family unit works together, so there is a certain degree of interdependence in religious affairs. Members pray for others' souls after death. In times of trial, members make promises to God for each other. They fulfill each other's promises if necessary and in general regard the family unit as responsible for obligations to the divine that any one member has incurred as it would be for any secular debt.

Except for certain wealthy households, rare exceptions, the family does not have a shrine. While there are pictures and little images in the home there are no tutelary deities, images of special devotion. For whatever reason, the family seems to be an exception to the rule that identities engender corresponding collective religious symbols.

It is difficult to say where the devotions of families leave off and the devotions of individuals begin. Individuals have saints or advocations of Mary to whom they are particularly devoted, but these are often passed down from parents. It is natural for a child to turn to the devotions with which his parents have had success, whose pictures are up around the house. Indeed, baptism itself, when the parent confers a saint's name on the child, often involves the transfer of a parent's devotion to the child. (Although it is by no means necessarily true that a person is especially devoted to the saint for whom he or she is named.)

Leaving aside those shrine devotions already discussed, how do these new devotions, with a family or an individual focus, enter the valley? What are they and how have they changed in past years? A number of sources will enable us to answer these questions with some degree of exactitude: The images in the valley churches, their rate of turnover and the way they are selected; names given to children at baptism; and the devotions most commonly propagated in devotional literature and missions. These all help provide the answers. The ensemble, together with the shrine images already discussed, forms the repertoire of devotions available to the family and the individual.

In very few churches do the original statues, designed to fit into the reredos, remain. Instead the statues in the churches are gradually changed, along with the tastes and affections of the times. Aside from the patronal images (active and formal), which change rarely if ever, the other images in the parish churches accumulate through the contributions of parishioners (usually the more wealthy ones), by the contribution of the priest or occasionally by diocesan mandate. They are removed by the priest, under the cover of a general renovation of the church or as a result of a loss of favor of a given devotion. Often this occurs less by active policy than by attrition. An image is destroyed by fire or simply wears out, and if it is no longer the object of devotion, that is, if it is no longer useful, it is simply not replaced.

Images have also been removed by the people. Spain has seen waves of overt anticlericalism (in addition to an ever-smouldering latent anticlericalism), and occasionally, as in the 1930's, these movements have included systematic iconoclasm. Several of the churches in the valley were swept clean of images during the Civil War. This provided the opportunity after the war to introduce new images and abandon some of the old ones. In all cases the desecration was done by militia from outside the valley, against the wishes of most villagers. In several villages in the valley the village *Comité*, or Republican junta, prevented the destruction of the village images, occasionally by threat of violence. It is impossible to say how much covert support the iconoclasts may have had in the valley. In many parishes the most prized images were taken out of the churches prior to the arrival of the militia and hidden until the arrival of the Nationalist troops in 1936. In San Sebastian some of the images were hidden in a winter barn. In Tudanca the image of Vado de la Reina was kept by an old lady in her fireplace. (Several priests were also hidden in these villages for longer or shorter periods until they could escape over the mountains into nationalist-held Castile.)

An incoming priest must work with the shrine images. He must adapt to them. On the other hand, most of the other images in the churches are devotions brought in by the priests or missionaries, each new generation bringing its favorites. The people, in turn, are the ones who must adjust to these additions. Priests who do not attend to the local devotions lose the favor of their parishioners. In the case of one Obeso priest who would not allow the statue of El Llano to leave the chapel for the annual folkdance ceremony and in the case of a priest of Tudanca who allowed the chapel of Vado de la Reina to collapse through neglect, the devotion survived them. The tendency to maintain devotions, sometimes in spite of a given priest, also occurs in the case of the priest-sponsored devotions. What one generation of priests has implanted with fervent sermons and devotional literature may be very hard for later priests to uproot, for a devotion continues in the parish with its own momentum, handed down within the family, although successive generations of priests may give it no special attention. Thus the devotions of the villagers, concretized in the images and pictures of the parish church, are like a series of geological strata, some dating back for centuries. Sometimes, after a lapse of several generations, devotions will peter out or be crowded out by an accumulation of new ones demanding time and attention from the devout, but often the image itself remains. Thus the mere listing of the images in the churches in the valley is not *ipso facto* evidence that the people are still paying attention to them. For this reason each generalized devotion must be considered separately, in order to see what stratum it comes from and whether it is still active.

Devotions come into and go out of style. In the church as in all cultural matters there are fads, which are brought in by the priests. The logic of replacement would seem to be the same on the level of parish images as that for shrines, an admixture of older devotions becoming intractable and unresponsive to the demands of the devotees and the attraction of novelty of new images. The ultimate origin of the trends that the parish priests bring to the valley has been the Papacy, which has encouraged different devotions according to the personal predilections of the Popes and the ever-present necessity to revitalize popular piety with novel devotions. The Roman Calendar changes, and the small parish churches of the Nansa valley, at the end of the long chain of command, follow suit. The changes are often sponsored by a religious order and sometimes are facilitated by a Pope from that order. The same order, on the local level, foments the devotion before and after it has become "legal."

Examples of the clear influence of orders upon local devotions are the devotion to the Holy Rosary, actively propagated by the Dominicans of Our Lady of Las Caldas through the establishment of brotherhoods; devotion to the Souls in Purgatory, encouraged by the Franciscans, who also established brotherhoods for this purpose; and the activities of the Carmelites in favor of Our Lady of Mt. Carmel. In the last century the institution of missions has resulted in the addition of many pictures or images in the churches: Our Lady of Perpetual Help by the Redemptorists, Passionist Saints by the Passionists, and Saint Anthony of Padua by the Capuchins (see Table 4).

TABLE 4

Images Most Common in the Churches and Chapels of the Nansa Valley (Polaciones to Camijanes) 1969[a,b]

Image	Number of different churches
1. Sacred Heart of Jesus	17
2. Immaculate Conception	11
2. Saint Joseph	11
4. Saint Anthony of Padua	10
5. Our Lady of Mt. Carmel	7
6. Our Lady of the Rosary	6
6. Our Lady of Fátima	6
6. Saint Roch	6
6. Crucifixion of Christ	6
10. Sacred Heart of Mary	5
11. Saint Michael	4
12. Saint Anthony the Abbott	3
12. Infant Jesus of Prague	3

[a]Number of images and pictures (per church): minimum 1; maximum 20; average 10.

[b]Ratio of images of Christ : Mary : saints for the entire valley is 1:2:2.

Information as to the potency of the new devotions may also come by word of mouth. More likely it comes in the form of propaganda distributed either by the missionaries themselves in the course of their mission or by magazines that they encourage the villagers to subscribe to. The most popular magazine, *El Santo*, is that dedicated to the devotion of Saint Anthony of Padua. It is published by the Capuchins of Santander. Subscribers also get a calendar with a suitable religious picture for their kitchens.

The devotion to Saint Anthony of Padua is of long standing in the valley. His feast was instituted in Spain during the reign of Philip V.[36] Aside from images of the Virgin, Saint Anthony of Padua is accorded more devotion than any other Saint. Men go to Saint Anthony (who has profited from the confusion with the patron of livestock, Saint Anthony the Abbot) for problems with animals and women go for help in matters of the heart and for lost objects. Along with the Virgin and Souls in Purgatory, in virtually every church there is an alms box for Saint Anthony.

Almost on a par with devotion to Saint Anthony is devotion to the Sacred Heart of Jesus. Every church in the valley has this statue. One wonders whether the diocese may have made its presence obligatory. The devotion is a recent one, dating from about 1910 in the valley. It began with a visionary nun in Paris in 1673. St. Margaret Mary Alacoque, a sister of the Order of the Visitation, had four visions of Jesus Christ from 1673 to 1675, "concerning devotion to his heart as symbolizing his love for mankind, which men so often reject."[37] The content of the visions was made known by her confessor, and the devotion was propagated in France by St. John Eudes. It was not accepted as a general feast of the Church, however, until 1856. It was at this time (1858) that Jesús began to appear as a given name in the baptismal records of the valley. Pope Leo XIII consecrated the world to the Sacred Heart of Jesus in 1899, and St. Margaret and St. John Eudes were canonized in 1920 and 1925, respectively. In 1915 an old chapel in Puentenansa dedicated to a shepherd saint was converted to be the chapel of the Sacred Heart of Jesus, and a spirited fiesta was held there annually (except for certain of the war years) until about 1960. On one occasion images were taken to the chapel from all the surrounding villages as a token of homage, just as the Church had organized the visits of images to the crowning of La Bien Aparecida. Several villages, including San Sebastian, had brotherhoods devoted to the Sacred Heart for a time. Under the dictatorship of Primo de Rivera, King Alfonso consecrated the nation to the Sacred Heart of Jesus, dedicating a national monument for this purpose at El Cerro de Los Angeles, outside Madrid. This monument, and the devotion itself, became a symbol for and of the Right in Spain. During the Civil War militiamen from Madrid would regularly go out to "execute" the monument, reducing it to rubble. Pictures of the ruined statue were widely distributed by the Nationalists, like the lithographs of a plane bombing the shrine of Our Lady of Pilar, as evidence of the depravity of the opposition. In the official Anarchist history of the war, published in Toulouse, a photograph of the destroyed monument is proudly included. Hence it is far from surprising that in the Nansa valley, as elsewhere in Spain, the Sacred Heart devotion stood for the Nationalist side during the war. After the war was over the statues that had been destroyed were immediately replaced by popular subscriptions organized by the Daughters of Maria.

Indeed, even before the Civil War, the Sacred Heart of Jesus had come to stand for the embattled struggle of the Church in the twentieth century against antichurch regimes. The pilgrimages in the Nansa valley to the Sacred Heart of Jesus bore some of this character—as a political and religious public witness of devotion, not so much to local nonbelievers, for they were few, but rather to the nation or the government.

The use of public religious procession as a *witness*, as a kind of defiant testimony, is common to all countries and all times when communities and nations are divided. Since this valley has for so long been united in its religious culture, its processions have not usually taken on such a note. But possibly the kind of politic-religious statement made in the processions in the 1930's and 1940's to the shrines to La Luz and the Sacred Heart of Jesus, were also made in the 1830's or during the Napoleonic invasions.

Women who grew up in the 1910-1940 period still maintain a devotion to the image. But it is not seen as a miraculous image, as Saint Anthony is. Saint Anthony appears holding the baby Jesus; perhaps he is seen as holding every mother's child, protecting every child. In earlier times, when infant mortality was high, the Saints with children must have been seen as the infants' protectors and their escorts after death. Saint Joseph and, of course, all the images of Mary except that of the Immaculate Conception could also be seen in this light. These Saints, the images of Mary with the Baby Jesus, and finally the statues of the Baby Jesus alone (Infant Jesus of Prague, The Baby Jesus Good Shepherd) epitomized and accentuated the cult of innocence and childhood maintained by the culture. The cult of childhood must have been related to the high infant mortality juxtaposed with the Church's emphasis upon the perils of the afterlife and the unpleasantness of purgatory. The Sacred Heart of Jesus, on the other hand, emphasized for the first time the adult, whole Jesus, not the child nor the crucified victim nor the patient suffering Nazarene carrying the Cross. In spite of its original emphasis upon redemptive love, it was the most fit symbol for a church in struggle. But this devotion has now waned considerably, perhaps because the new Church has reached some accommodations with the modern world.

We know from the establishment of chapels in Obeso and Tudanca that statues of the Immaculate Conception date at least as far back as 1620 in the valley. Now they are variously called "La Inmaculada" or "La Purisma." The great majority of the churches have this image. Perhaps some added it after the

Papal definition of the dogma in 1854, but most seem to date from before that, though no statue seems older than the seventeenth century. The devotion probably entered this valley with the theological controversy in the eighteenth century, brought, at least to Obeso, by a man who spent most of his life in Andalusia, where Franciscans (with immense popular support) and Dominicans battled out the validity or invalidity of the doctrine, respectively.[38] The popularity of the devotion probably increased following the apparitions at Lourdes. Of all images of Mary, this is the one that most emphasizes her purity (the dogma has to do with the purity of her conception in St. Ann), and so it became the symbol for the organizations such as the Daughters of Maria that were dedicated to the maintenance of a chaste and sober life while in the unmarried state. I have no evidence that the devotion was "pushed" by any particular order, although the Franciscans would seem to be appropriate candidates.

Of the other devotions to aspects of Mary, the most important are, of course, the shrine images alluded to at some length above, whether in chapels or parish churches. Devotion to Our Lady of Mt. Carmel is quite intense except, it would seem, in Polaciones, which does not have this image. Devotion to the Immaculate Heart of Mary seems to date in the valley from after the Civil War for it was not made a universal feast of the Church until 1944. Already the popularity of the devotion has waned.

Another recent devotion is that of Our Lady of Fátima, one that also dates in this valley from after the war. In 1949 the image of Fátima toured many dioceses in Spain, including Santander, to arouse devotion, with much fanfare.[39] There are six images in churches in the valley, and several churches have alms boxes for the Fátima devotion. Because of the content of the apparition message (the conversion of Russia) and the political context (a radical regime in Portugal), the Fátima apparition became a weapon in the Cold War, especially in the 1950's, as the devotion to the Sacred Heart had been before. While I was there, however, it seemed that the devotion had peaked, probably because of the disappointment in 1960 when the third and last secret message was supposed to be revealed, and never was.[40] In any case the Fátima story has been somewhat superseded.

Saint Joseph, whose relative popularity seems to be due to his title of "patron of the good death," is to be found in about two out of three churches. In most cases he has survived the postwar renovations, but I never heard anyone refer to him as a member of their personal pantheon and he is no longer honored ceremonially. Villages used to celebrate his fiesta, but that fiesta was abolished by the Church, which may explain his relative lack of importance at the present time.

Indeed, few if any of the other saints besides the Saint Anthonys receive devotion today. In the past there seems to have been devotion to Saint Michael. The parish church of Cosío was dedicated to him, and all of the surrounding villages had his image. The great fair of Rionansa is held on his day, a day of rest, relaxation, and recreation that is looked forward to all summer.[41]

In the nearby territory of Campóo de Suso there was a chapel dedicated to Saint Michael that was changed in the seventeenth century to Our Lady of the Snows. Similarly it appears that the older special devotion to Saint Michael has been superseded in the Nansa valley by one or another image of the Virgin. My guess is that, as a warrior saint, he had a certain appeal during the reconquest of Spain. We saw before that many persons in the valley took part in the reconquest. These factors might help explain the relative irrelevance of the devotion as the communities settled down to a peaceful existence. A similar explanation may hold both for the existence and relative lack of devotion to Saint George, the titular patron of the Puentenansa church, and Saint Sebastian, another soldier saint, in the San Sebastian church.

Another saint to fall from favor has been Saint Roch, a pilgrim saint shown always with a dog, often also with a child, who was a protector against disease. Copatron of two widely separated villages (Carmona and St. Eulalia in Polaciones), there were subchapels dedicated to him in many other villages. St. Roch is by no means as ancient a devotion as Saint Michael. He acquired special popularity in the fifteenth-eighteenth centuries. Recently the Vatican has removed him from the Roman calendar, as grave doubts to his authenticity have arisen. But in any case his usefulness as a protector from plague and other epidemics had already come to an end some time ago.

The saints in the parish churches that occur in the valley with the least frequency are the titular patrons. Out of the 23 dedications in the valley, 16 are to saints that do not appear even as images in any of the other churches. This must be because the dedication of the church occurred at its founding, and the popularity of the devotion was not a problem. Some priest or bishop simply made a decision. A crucial element in that decision was that the patron of the church be different from that of the surrounding churches in order to distinguish the parishes by endowing each village with a different festival day and a different symbol to rally around. Hence in the entire valley there are only four duplications of dedications of parish church titular images, and of these three are between villages of Polaciones, cut off in another Diocese until 1949, and the rest of the valley.

Churches accumulate images. It is easier to add them than to take them away. The priest is only legally in control of the parish church. He must live with the parishioners. Only recently, with an emphasis upon medieval restorations, has there been a movement away from images for artistic reasons, and only two churches in the valley, those of Obeso and Celis, have gone in this direction. In Obeso it was the decision of the priest who in other respects is well liked by the parishioners. The villagers would have preferred to keep the old saints, but the priest said that a simple crucifix was "more modern." "And as we don't know about these things, and as he gives the orders, that's what happened," they told me. The relative lack of opposition that the priest encountered is a measure of the extent to which the Church's emphasis upon Mary and Christ and the deemphasis of the saints in the past 40 or 50 years has influenced parish devotions. Unless the occasion for a renovation occurs, the younger priests see no reason to remove the images, even though, for the most part, they neither understand nor respect the old-time devotions. The older priests, even if they no longer actively encourage mediation of communication to God through the saints, see nothing wrong with it.

As for the people, many of them regard the images as familiar parts of their church, part of the village's cultural heritage: they see no reason to remove them. Aside from devotion to Saint Anthony, they hold no communication with most of the saints in their church. In San Sebastian there used to be collections of maize every year for 10 saints that used to be in the church. Some of the images, after the war, are no longer there. Others have no appeal; the custom is dying out.[42] The saints, it now seems, have little to offer. Their specialties have been superseded by human specialists (doctors, pharmacists, veterinarians) or divine generalists (Christ, the Virgin). Those Saints that gave protection against the weather are not so necessary now, as the relative economic well-being makes any particular harvest less critical. The anxiety that centered particularly on the mies has subsided, since the villagers now buy their bread.

VIII. Baptismal Names

Many devotions were carried on in the home as well as the churches and chapels.[43] One measure of the strength of these devotions over time is the use of saints' names at baptism. The parish records of San Sebastian contain all baptisms in the village from 1638 to the present. I copied the names and calculated the frequency of the most commonly used names for 10-year periods over the three centuries as one more measure of the amount of devotion to the different saints. Two factors complicate the matter. One is the tendency to choose the name merely from the calendar—the saint of the day of birth. The other is to give the name of the father, mother, grandparent, or godparent. The former situation is easy to detect, and seems to have occurred especially in the years 1808 to 1940, but not before and little since. The naming after parents, etc., on the other hand, does not preclude a devotion to the saint named, and evidence from the listing of relatives' names in the records indicates that such naming is infrequent. Finally, it is possible that in very recent years radio and television have been providing new names from the names of recording artists and other public personalities. Table 5 shows the rank order of frequency of saints' names that appear with any degree of frequency (above 5 percent of total names given in 1638-1807, above 2 percent in 1858-1967) in the baptismal

TABLE 5

Rank Order Frequency of Saints' Names, 1638-1967, San Sebastian

Name	Rank Order			
	1908-1967	1858-1907	1708-1807	1638-1707
Maria	1	1	1	1
José, Josefa	2	2	5	–
Antonio, -a	3	–	4	–
Manuel, -a	4	6	6	–
Jesús, -a	4	–	–	–
Francisco, -a	4	3	2	2
Pedro, Petra	–	4	–	–
Juan, -a	–	5	3	3
Domingo, -a	–	–	–	4
Catalina	–	–	–	5

records of San Sebastian. The years 1808-1858 are not given because in these years names are almost exclusively given on the basis of birth date.[44]

Table 5 shows that aside from the name Maria a whole new set of saints' names has gained popularity since the seventeenth century. Perhaps some of these changes reflect the differential popularity, indeed, the differential activity, of various orders in and near the valley. This might be the case with the decline of Francisco and Domingo, corresponding to the abandonment or restriction of activity of the nearby monasteries of Franciscans (at San Vicente) and Dominicans (Las Caldas) in the nineteenth century. Conversely, the renewed popularity of Antonio may be due to the influence of the Capuchins of Santander and their journal.

The idea that names bear a relation to devotions is bolstered by the cases of those saints or devotions to which a definite date can be independently assigned. Hence the name Jesus appears first at the entry of the feast into the Roman calendar (1856) but remains scarce until the cult was actively spread in the valley, in the first years of the present century. Similarly the name Ramón suddenly began to be used in San Sebastian in 1742, at the time a chapel was established there. The rise of popularity of the name José corresponds to the spread of his devotions in the manuals of the last two centuries. This rise began about 1708 and has continued, particularly marked in the past three decades.

Table 6 gives the percentages upon which Table 5 is based. The percentages given are averages; the averages of the percentages of occurrences, calculated by 10-year intervals, of a given name out of all the names (not out of all the people, as there are more names than people).[45]

The figures from 1858 to the present reflect the very general drop-off of devotion to saints in general, with a concomitant increase in veneration of Mary

TABLE 6

San Sebastian: Average of the 10-Year Frequencies (in Percent) of Names

Name	1908-1967	1858-1907	1708-1807	1638-1707
Maria[a]	33.5%	11.6%	25.0%	41.0%
José, -fa	6.8	3.6	5.7	1.1
Antonio, -a	3.2	1.0	6.4	4.3
Manuel, -a	2.7	2.0	5.4	0.9
Jesús, -a	2.7	0.8	0.0	0.0
Francisco, -a	1.3	3.0	10.9	16.0
Pedro, Petra	1.2	2.4	2.1	1.5
Juan, -a	1.0	2.2	9.3	9.6
Domingo, -a	0.2	0.9	3.5	7.3
Catalina[a]	0.0	0.0	1.3	8.3

[a]Because these names have no masculine cognates, the percentages are of total names given to females.

and, to a lesser extent, Jesus, that we discussed above. Nowadays, when I ask why the saints are not more venerated, I am told by the villagers that they go "to the main doors, not the side doors"—to the Virgin and to Christ, not to the less powerful saints. The only exceptions, it would seem are Saint Anthony and to a very much lesser extent, Saint Joseph.

Similar figures from the village and township of Tudanca from 1858 to the present confirm the evolution of devotion as manifest in the San Sebastian records. The major lines are the same. The surge of the use of Mary follows the San Sebastian record step by step (see Fig. 8). There is a parallel decline in the use of Pedro and Francisco and a similar growth in the popularity of Antonio, José, and Jesús. In the two periods 1858-1907 and 1908-1967, the two towns share four out of the five most commonly given names. Differences largely involve the less popular names: Tudanca seems to have more affection for the names Miguel, Angel, and Teresa. But the general lines are the same.

That the two patterns are similar should not be surprising, for the extent to which, on this level of generalized devotions, each village has its own religious culture can easily be overemphasized. They are all participating in a diocesan-wide, nationwide, Europewide Catholic society in which saints gain and lose popularity much as secular heroes do. As in the case of secular heroes, much seems to depend on those individuals and institutions that propagate and mediate information concerning the saint. The orders, the diocese, the seminary, devotional books and pamphlets, and indulgences granted by the Pope all play their role in the popularity of these generalized, nonlocal devotions, and we should not expect great idiosyncrasies at the village level. Indeed, the fact that two villages have congruent figures encourages us to think that the giving of

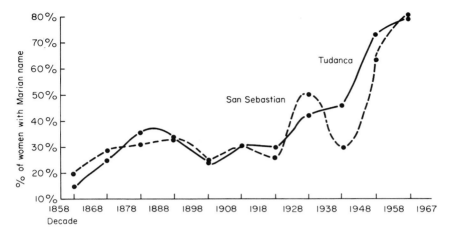

Fig. 8. Graph of the percentage of women named for Mary, 1858-1967, two villages.

names does indeed have something to do with devotion, as opposed to secular family traditions. And the specific names that have gained popularity point more toward Church doctrine and propaganda than toward mere fads, as with names in a more secular society.

The increase in the use of Mary as a name in the nineteenth and twentieth centuries was accompanied by a diversification of Marian titles as baptismal names. The earliest mention of a particular kind of Mary in a given name was 1768, for San Sebastian, when a child was named Guadalupe. The practice began in earnest in the 1880's and today there are few daughters simply named Mary. The overwhelming proportion are called by some variety of Mary. Table 7 is a tabulation of baptismal names devoted to the different varieties, or advocations, in San Sebastian and Tudanca.

If these names were being given simply on the basis of date of birth the expected frequency for any name, assuming that the birth fell randomly during the year, would be 3.5 over the time period in question. (There were 1265 female babies born in these towns from 1768 to the present; 1265 × 1/365 = 3.5.) Hence those listed below Table 7 could be explained away in this manner, but those in the table, especially those toward the top of the table, must be determined by some sort of preference other than birth date. Since Our Lady of Mt. Carmel and the Immaculate Conception are at the top of the list and are two attributes of Mary for which there are the most images in the valley, we are safe in concluding that there is a measure of piety operating in the choice of other attributes.

The list points to several devotions not present in the parish churches. Our Lady of the Sorrows (Dolores) spread throughout Mediterranean Europe in the

TABLE 7

Advocations of Mary in Baptismal Names, San Sebastian and Tudanca[a]

Name	Cases	Period
1. Carmen	47	Especially since 1930
2. Dolores	30	Two periods: pre-1900; and postwar
3. Inmaculada Concepción	29	
4. Pilar	19	
5. Mercedes	18	
6. Angeles	16	twentieth century, postwar
7. Asunción	14	Few lately
8. Rosario	14	Post-1880
9. Nieves	13	Especially Tudanca (chapel there)
10. Milagros	11	More recent
11. Remedios	10	Especially San Sebastian (minichapel there)
12. Luz	9	All twentieth century (chapel Peña Sagra)

[a]The others, in order of frequency, are: Guadalupe, Visitación (8); Soledad (7); Consolación (6); Encarnación, Purificatión (=Candelaria) (5); Virtudes, Paz, Amparo (3); Socorro, Cruz, Patrocino, Mar, Lourdes, Rocío, Montserrat, Covadonga, Loreto (2); Bien Aparecida, Piedad, Sagrario, Araceli (1).

fifteenth century. The devotion is above all to the anguish of Mary at the crucifixion, but includes other sorrows in her life as well. The Servite order propagated a special devotion to the Seven Sorrows of Mary. The Saint's days for Dolores are the Friday after Passion Sunday and September 15. The only other overt devotion to the Sorrowing Mother that I have come across in the valley, apart from the name, is the wearing of habits *(vide supra)*. Those who are wearing colored habits, for instance the habit of Mt. Carmel, switch them for a black habit, that of Our Lady of Sorrows, when they go into mourning.

The devotion to Our Lady of Ransom (Mercedes) was propagated by the Mercedarian order, founded for the rescue of Christian hostages captured by North African pirates. The feast date, at first limited to the order itself, was extended to cover all of Spain by Innocent X (1644-1655). An image came into the valley at least as early as 1742, when the shrine to San Ramón was founded in San Sebastian.

I do not know about the origin or spread of the devotion to Our Lady of the Angels (Los Angeles), except that it was a Franciscan devotion dating back to the time of St. Francis himself. The feast is not celebrated in the valley, there are no images of the Virgin, and no other overt sign of devotion.

The Assumption of Mary (August 15) was previously the major feast day of the Virgin, supplanting the Annunciation (March 25). Two Marian feasts are celebrated on this day in the valley, that of Our Lady of the Lowlands, of

Obeso, and Our Lady of the Vega, of Santotís. This day is sometimes referred to as Santa Maria, or Santa Maria de Agosto, to distinguish the day from September 8, the Nativity of the Virgin, known as Nuestra Señora. In the past 400 years September 8 has become the Marian feast day most preferred, and some shrines have changed their fiesta day from the Assumption to the Nativity.

The devotion to Our Lady of the Miracles (Milagros, La Milagrosa) is connected with a small traveling image that circulates through the households in each village. This practice was begun in this valley in the 1920's and is still actively maintained. The small image of the Virgin travels from house to house in a wooden traveling case. The case may be set up in the kitchen or in a bedroom, and usually the lady of the house will light a candle or a little oil lamp in front of the image. The box has a slot in it for coins, which are used to purchase masses in the church. While the practice was certainly once set up by the parish priest or missionary, it has taken on a rhythm and life of its own. It is a special way that the divine presence penetrates the home. As it makes a circuit in the village, it also symbolizes the concerns that the villagers have in common. There is also a shrine to Our Lady of the Miracles in the village of Torres, near Torrelavega, and a statue in the school run by nuns in Cabuérniga, and these too may be sources for the use of the name. The devotion ultimately originated in a series of apparitions of the Virgin to a nun, Catherine Labouré, in Paris in 1830.

IX. Brotherhoods and Missions

The Franciscans and the Dominicans no longer actively influence the devotions of the valley, but they have left behind their legacies: the devotions to souls in purgatory and the rosary. The devotion to the souls in purgatory *(las Benditas Animas)* is neither concretized in an image nor mentioned in baptismal names, yet is very strong in this region. In the list of active patrons (p. 68) three villages are listed as having no active patron—San Sebastian, Sarceda, and Rozadío. In all three villages there has been an inordinate amount of devotion to the souls in purgatory. In all of the parish churches the small side altar to the left of the main altar has been dedicated to the souls in purgatory.

Devotion to souls in purgatory is related, of course, to the villagers' concern for the dead. During the week following a person's death, on the anniversary of the death, and on the deceased's saint's day, masses are said for the repose of the soul. And the candles burned for years after a person's death were burned especially beside the altar dedicated to the souls in purgatory. This custom survives in Tudanca, but in few of the other churches. In Tudanca the candles are lit by the women before they attend any service; they are extinguished when they leave. This kind of cult is centered in the church rather than the cemetery because until the beginning of this century, the dead were buried in the church itself. It is considered that the living owe the dead the prayers that would put them in repose. Local folklore abounds in stories about ghosts and graveyards, the souls in purgatory being envisioned as white, shrouded objects, sometimes with chains. Particularly in Sarceda there have been recent cases of the living accompanied on trips up the mountain by the dead, usually at night. This kind of devotion for and worry about the dead is characteristic of the entire Atlantic fringe from Galicia through Brittany, Ireland, and England.

The devotion to souls in purgatory is a separate matter from devotion for specific ancestors. For the uninitiated, purgatory is "the place and state in which souls suffer for a while and are purged after death, before they go to Heaven, on account of their sins."[46] The sins in question are unrepented venial sins or imperfectly repented mortal sins. If a person has committed completely unrepented mortal sins, he goes to Hell. Purgatory, on the other hand, is a waystation to Heaven. Souls in purgatory can have their stay shortened by special devotions that they themselves performed while alive, also by the devotions of persons still living who pray for them or who apply indulgences to them. It has also long been the custom, although unsanctioned by official Church policy, for the faithful to pray *to* the souls in purgatory, just as they would to any saint. For this reason in the parish church of San Sebastian there is an alms box for the *animas* in which the villagers place thank-offerings for services rendered in response to petitions.

A young priest in San Sebastian once complained to me that people have the false notion that they can pray to the souls in purgatory, instead of for them. The notion, however, is one inculcated by his own predecessors. As in many other cases, devotional practices which the clergy now consider to be folkloric or deviationist are often in fact the practices taught by representatives of their Church two or three generations before. Much of what passes in anthropology as the "great tradition" and the "little tradition" are in fact just simultaneously extant practices that date from different periods of the same institution. In the case of praying to the souls in purgatory, a devotional handbook composed in the early nineteenth century, which by 1945 had run to 174 editions, and which was commonly used in the valley explains why it is possible to pray to souls in purgatory and why such practices will be rewarded:

In no case of necessity is gratitude so sure as from the souls in purgatory. In this life the bad are almost always ungrateful, and the good are capable of ingratitude as they can always go bad. But those souls cannot help but be most grateful because they cannot help but become saintly. For this reason they cry out incessantly for their benefactors, and the Lord pays attention to them because they are in his grace; and they will cry out even more and will be better heard when they ascend into heaven. And as the favor that is done for them by accelerating their progress towards glory is beyond comprehension, so the efficacy with which they cry to God for their benefactors is imponderable.[47]

In another passage, referring to a novena for the animas, the same handbook states:

> This novena can be done at any time of the year, and it will be very appropriate to do it when any special favor is desired from the Lord, whether it is for oneself or for somebody else; because it is a very efficacious means of putting God in one's debt to make this spiritual alms to His imprisoned and afflicted spouses.[48]

The epoch of origin of the devotion in the valley is impossible to ascertain. Masses were being willed to the souls in purgatory when the death records began to be kept in 1615, so the practice is at least more than 350 years old in the valley. It became somewhat organized with the establishment of brotherhoods in the late seventeenth century. Brotherhoods were set up in Sarceda, Obeso, and Tudanca in 1678, and probably in other villages as well.

The account book of the brotherhood of Sarceda has been preserved with the original rules intact. It states that a Franciscan was in the village as a missionary on March 13, 1678; on that day he proposed the establishment of a brotherhood, as he had already done in other parishes. He proceeded to name local officers for the brotherhood: an Abbot, two majordomos, a treasurer, an accountant, a secretary, and an alms-seeker. These posts involved the solicitation and administration of money and grain collected for the purchase of candles and the payment of masses to be said for the souls in purgatory and the ringing of a bell every evening in the alleys to remind people to say a prayer for the dead. All members of the brotherhood were to gather for four masses a year, hold a general meeting every year on Saint Joseph's day to elect the officers for the next year, and hold a service and procession at vespers on the first Sunday of every month (men first, then clergy, and at the end, women, "so there is no confusion").[49]

The account book was maintained quite faithfully, with 25 lists of new members in the eighteenth century and 13 lists of new members in the nineteenth century. The accounts, which kept track of lands deeded to the brotherhood and the income from the advance sale of the hay on the lands every April, were kept regularly until 1915, with one entry in 1930. That of Obeso was maintained until 1937. In Sarceda in 1896 there were 14 fields belonging to the souls in purgatory, four more belonging to Our Lady of the Rosary. It appears from supplementary regulations added in 1781 to the brotherhood rules that the major thrust of the brotherhood was the well-being of the souls of the brothers on their demise, indeed, the brotherhood acted as a burial association as well as a prayer collective.

It is probable that the intense cult of the souls in purgatory, which seems to have continued through the nineteenth century, played a contributing role to the popularity of Our Lady of Mt. Carmel. The scapular was offered as a way out of purgatory. The sabatine indulgence, spuriously attributed to Pope John XXII (1322) but later confirmed by Pope Paul V (1602-1621), established that those wearing the scapular of Our Lady of Mt. Carmel and fulfilling the conditions of its use would be automatically delivered from purgatory the Saturday after their death. I think it likely that the devotion and attention to souls in purgatory in the seventeenth and eighteenth centuries in this valley paved the way for the flourishing of the devotion to Our Lady of Mt. Carmel when it was propagated by the Carmelite orders, for the devotion to Our Lady of Mt. Carmel is the Marian response to purgatory *par excellence.*[50]

Brotherhoods of the Rosary were set up in the valley with conditions very similar to those of the brotherhoods to the souls in purgatory. Those in the valley came under the jurisdiction of the Dominicans of Las Caldas and were set up, or re-set up, in the seventeenth, eighteenth, and late nineteenth centuries.[51]

Devotion to the rosary, even without the establishment of brotherhoods, was generally encouraged by the bishops. Time and again mention is made in the Episcopal visits to the parish recorded in parish account books of the virtue and necessity of the communal recitation of the rosary. The devotion was encouraged by a number of Popes in the seventeenth and eighteenth centuries subsequent to its official establishment as a holy day in 1573 by Gregory XIII. Pastoral letters from the Bishop of Burgos recommend the daily recitation of the rosary in the church in 1699, and in 1704 there was an order that it be recited on all feast days after mass and on regular work days at nightfall. This order was repeated in 1731 accompanied by indulgences and was reaffirmed by the Bishop of Santander in 1787.

There seems to have been a general lapse in the activity of the brotherhoods in the first part of the nineteenth century, perhaps as a result of the exclaustration of the Dominicans, but the devotion of the rosary revived at the turn of the century. The revival was partly due to the activities of Pope Leo XIII in its favor. The parish priest of Sarceda notes with satisfaction in 1912 that all members of the brotherhood (he listed 32 men and 52 women) received communion on the feast day of the Rosary, October 7.

The rules of the brotherhood were simple. One had to recite one complete rosary (15 sets of 10 Hail Mary's, separated by the Lord's Prayer) each week. In return one would share in the spiritual benefits gained by all the prayers of all the members of the brotherhood in the world. In addition one was required to attend four anniversary services per year for deceased members, attend the funeral of brother members, and process around the village with rosaries and a lighted candle in hand on the first Sunday of every month. Members were given the right to inscribe deceased persons as members, thereby gaining for them the privileges and indulgences earned by all members for each other. The brotherhoods were prayer cooperatives that enabled one to amplify one's prayer power many thousandfold. In the Brotherhood of the Rosary the members were allowed to share not only each other's spiritual earnings, but also those of all members of the Dominican order.

In recent times the rosary has become so widely accepted that it is hardly a separate advocation, a separate attribute of the Virgin Mary. The apparitions at Lourdes, Fátima, and San Sebastian, all of which have engendered separate devotions, all emphasized the devotion of the rosary.

Indeed, with all of these Marian devotions it would be misleading to overemphasize the aspect of competition. Certainly different orders, with different advocations of Mary, are constantly putting forward different devotions. But when considering the different generalized devotions popular with individuals now or in earlier days it should be remembered that at no point are the different devotions incompatible with one another. Indeed, the devotional handbooks that are still used by the devout women in the villages are compendia of the different devotions. The orders and the other outside sources from the Church shaped the repertoire of generalized devotions available to the villagers, but usually their devotions were mediated through the parish church and the parish priest. He called in the missions; he was responsible for the parish church images; he was the key mover in the brotherhoods. As a result, no particular order could long predominate, and the villagers obtained an amalgam of devotions.

Take for instance the contents of the handbook written by St. Anthony Maria Claret in the nineteenth century and augmented thereafter. The 174th edition, which I found in the house where I stayed in San Sebastian, included the following devotions. (To demonstrate how accurately this handbook reflects the generalized devotions of the valley, I have starred those devotions which have been shown in this chapter to be particularly common.):

> Devotion to the Holy Trinity
> *Devotion to the Heart of Jesus
> *Rosary in honor of the Most Holy Virgin
> *Crown of the Seven Sorrows of Mary
> Scapular of the Heart of Mary
> *Scapular of the Virgin of Mt. Carmel
> *Scapular of the Immaculate Conception
> Devotion to the Heart of Mary
> Holy Exercise of the Stations of the Cross
> *Devotion to Saint Joseph
> *Devotion to Saint Anthony of Padua
> Saint Raphael, or the solace of the sick
> *Novena for the Holy Souls in Purgatory
> Reflections on the Passion of Our Lord Jesus Christ

It should be emphasized that each individual, each family, has a battery of these devotions and by no means limit themselves to one of them.

The impact of the orders has continued in this century, and missions are still being held. The orders involved, however, have changed. The monasteries are no longer centers for far-flung trips of evangelization.

Almost every village church has a big mission cross inside upon which is inscribed the name of the order that held the mission and the date it took place. It serves as a reminder of the lessons learned during the mission. For a mission is in a real sense a revival. It generally involves an intensive weekend of sermons, confessions, and communions. It is usually during the winter, when the villagers have less work. Some of those that have taken place in the valley have been run by Redemptorists (San Sebastian, 1940, 1951; Carmona, 1950; Cosío, 1956; Tudanca, 1956), Capuchins (Tudanca, 1965) and Passionists (Bielba, 1953).

The parish priest is too familiar a figure to get the people stirred up or to put the fear of death and fear of God into his parish. So an outside speaker is called in from time to time to rouse the parishoners from what the Church considers to be their apathy. Sometimes a pious parishoner will leave money for the expense. Missionaries have traditionally emphasized the perils of hell and purgatory and the need for repentance. The following is the text of a souvenir from the Tudanca mission of 1956:

> *If you wish to save yourself:*
>   Obey the commandments.
> *All the days of my life:*
>   I will pray three Hail Marys.
> *Every week:*
>   I will hear mass Sundays and Holy days.
> *Every year:*
>   I will confess and receive communion.
>   I will frequent the sacraments, doing
>   the nine first Fridays.
>
> "The devotee of Mary will not be condemned."
> (St. Alfonso Maria de Ligorio)
>
>   Tudanca, 29/1/1956

In spite of the somberness of the preaching, a mission is something of a holiday, a break in the dreary winter routine. People come from all the surrounding villages to hear the preachers, who are usually very good speakers. In the old days it was one of the rare opportunities, like the great fair, for women to go to another village, and it was eagerly looked forward to.

Along with their message of spiritual renewal and attention to the sacraments the missionaries bring the devotion of their order: little cards with prayers and pictures, large lithographs or even statues to be placed in the parish church, and subscription forms for the magazine of the order. The purpose is hardly mercenary. It is rather an honest attempt to spread a devotion in which the missionaries believe: the Capuchins, Saint Anthony: the Passionists, the Passion of Christ and the Sorrowing Mother; and the Redemptorists, Our Lady of Perpetual Help. It was suggested that the change in popularity of names from the seventeeth to the twentieth century had to do with the falling off of the Franciscan and Dominican missions and the succession of other orders in their place. Once foci of enormous spiritual power exercising a pervasive influence upon the villages in their districts, possibly because of their supposed efficacy as auxiliaries at the time of death, the monasteries were struck a blow by the exclaustrations of the nineteenth century, a blow from which they have yet fully to recover. Their passing has left a vaccuum that has been partially filled by more mobile religious elements, not necessarily based in the countryside. The result has been a diversification, a proliferation of devotions, of medals, scapulars, gimcracks, images, and magazines, the overwhelming percentage of which centered upon some form of Mary.

In the uphill battle against secularization in western Europe after 1800, the Virgin Mary, literally or figuratively, depending on one's point of view, led the way through a series of apparitions that reaffirmed the power and the usefulness of the supernatural in an age of increasing mechanization and materialism. There resulted, for the purposes of such a great battle on such a broad front, a series of generalized devotions, that on the level of the individual and the family came to supplement the fixed local devotions of the valley shrines, which had been dedicated to Mary at an earlier stage. The purveyors of these generalized devotions were themselves orders not participating in the mysteries of the sacred landscape on the local level.

X. Summary

The order of presentation of the devotions of the Nansa valley corresponds to the order of presentation of the general ethnography in the first section. As far as the shrines were concerned, there were some rough equivalencies: Certain shrines seemed to correspond fairly well with certain levels of identity, indeed, they seemed to be utilized as symbols for that identity. This was especially true, of course, for shrines organized to that end—some of the national shrines, the shrine that was shaped to fit the new provincial identity; a couple of shrines that had come to stand for vales; and, above all the shrines that stood for villages. One equivalence seemed clear: The more clear-cut the sociogeographic unit, the sharper its boundaries, then the more likely it was to have a cultural symbol, a protector, a patron in a shrine image. The jurisdictions, the territories of grace of the images in the valley and the region, could be no more clear than the boundaries of the sociogeographical units themselves.

The last half of the section dealt with the free-floating devotions which apply above all to families and individuals. The phenomena described in the two halves of the section stand in sharp contrast. The chapels for the most part date back at least 400 years, probably much farther. They do not budge. They are the cultural contribution of the valley to the repertoire of devotion. They may have been set up by someone at some time, but whoever or whatever organization it was is long gone or totally changed. The shrines remain. The legends are still remembered. The devotional practices associated with them go on. As symbols for social identities, as measures of belonging, as cultural boundary markers, they have a function that is primeval. They are virtually totem objects, embodying in some way the essence of the humanity of their devotees. I think of the tears, the great collective wrenching that took place in Sopeña, over the hill from Rionansa in Cabuérniga, when their image of Our Lady of Mt. Carmel left the village for the first time to go to the coronation of La Bien Aparecida in Santander. The role of active patron is deep and profound: it has a hold on the heart: it should not be underestimated or lightly dismissed. Its power can be measured precisely by the violence with which images of cherished devotions were pursued and, if possible, destroyed during the Civil War. The images literally embody a whole set of cultural values, so much so that they were considered by some to impede critically the revolution. Gide uses in his journal the image of a chicken having to break his shell in order to be free and applies this to the religious persecutions of the Republicans. The degree to which the images were destroyed is a measure of the degree to which they constituted a kind of cultural shell, a nexus of values.

I doubt that an intuitive sense of this affection and its converse antagonism can adequately be conveyed to those who have not experienced the culture. When I search for equivalencies in America the closest I can come is the sense of reverence which some people have for their flag or their capitol building, and its converse, a need to desecrate them. H. G. Wells talks of cultures with male symbols as communities of will and those with female symbols as communities of obedience: therein, perhaps, lies a reason why the flag is such a poor analogy to the Virgin, and why it is so difficult to transpose oneself from America to Spain and feel what the Spaniards feel about the Virgin as a collective symbol. Some Americans honor the flag and consider it sacred, but the sentiment toward these shrine images in Spain is an intense mixture of honor, reverence, and several varieties of love. It is an emotion predicated upon a sentient object, something that responds. When the Virgin is the image in question, the closest way to describe the emotion is that of the love for one's mother. But not the mother in a nuclear family, rather the mother in the extended family, described in the first chapter: the pivot, the fulcrum, the hub of the social relations of many, many people. In Spain the Virgin has become the pivot, the fulcrum, the hub of the emotional and cultural relations of whole collectivities.

In contrast to this type of role that the Virgin plays—as the epitome of a person's social self, as a symbol of one or another of a person's sociogeographical identities—there are roles that the Virgin and other saints play for the person *qua* individual. Many of the very same chapel or shrine devotions fill this role, especially in those villages, like Obeso, Carmona, Bielba, and Tudanca, where there are strong images in the shrines. But even in those villages the chapel images are supplemented by a mass of generalized devotions that have come from outside the valley cultural system, imports of one ilk or another, generally in the last 200 years.[52] The impetus of these devotions, as propagated by the monasteries and the mendicant orders, has been to encourage a personal and family interest in salvation. They are very much oriented toward the hereafter, in contrast to the shrine devotions, which are more concerned with the regulation of life and society in this world. The shrines, generally without the intervention of the priests, dispense mana. The generalized devotions are auxiliaries of the priests: they dispense absolution and salvation.

The shrines are energy transformation stations—the loci for the transformation of divine energy for human purposes and the transformation of human energy for divine purposes. The generalized devotions have to do with the transformation of people from one condition to another. The religion of any individual in the valley is a mixture of the two kinds of devotions discussed in this section. The two systems have been adulterated by each other. Generalized devotions such as Mt. Carmel have been appropriated for use as shrine images, and the shrines have been endowed with indulgences used to stimulate the same kind of personal salvationary religion that characterizes the generalized devotions.

The way the devotions are used, the modalities and laws governing the relations of individuals with God, by way of divine intermediaries like the Virgin, will be discussed in the following chapter. To a certain extent we will be intruding on their private lives, to an extent not at all present in our discussion of the public, collective devotions surveyed above. Our only excuse for such behavior is that we are not finding out about others so much as we are learning about ourselves, their fellows. In this chapter we have enumerated the divine figures that make up the pantheon of the valley, and we have seen the extent to which they permeate the landscape inside and outside of the villages. Then given the social setting and the divine repertoire, the next chapter will examine how different types of people make contact with God.

The village of San Sebastian de Garabandal.

Tudanca in February. The church and the school are in the foreground.

Peña Sagra and La Lastra from Tudanca, winter 1972.

After mass at Tudanca, February 1972.

Watching television in the tavern of Tudanca.

The annual outdoor mass at Nuestra Señora del Brezo, September 21, 1969.

The Virgin presides over the nine concelebrants at El Brezo.

The shrine of Nuestra Señora del Llano, with Obeso and Hoz Alba, the site of the apparition, in the background.

Recital of the Rosary during the annual "subida" of our Lady of the Snows from Tudanca to the shrine in the mountains, July 25.

The altar of the shrine of Nuestra Señora del Llano.

A relic of past devotions; what used to be a barrio chapel dedicated to the Immaculate Conception, in Obeso.

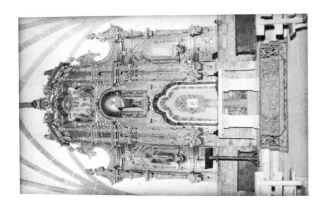

The saints in a typical valley parish church, (Tudanca).

An altar without saints — the newly remodeled parish church of Obeso.

Two herdsmen from Obeso in the upper meadows at the foot of Peña Sagra in May, 1969. Three years later both had abandoned traditional herding for good; one to tend a small number of milk cows in Obeso village; the other to work in Santander.

San Sebastian, May, 1969.

CHAPTER THREE

PERSON AND GOD

An understanding of how individuals and communities use the divine repertoire will be furthered by a summary of the kinds of communication with the divine that are common to the valley. These communications vary from quite generalized, affective messages to very specific, instrumental requests:

1. Generalized affective prayers
2. Prayers for the fulfillment of the annual round
3. Prayers for forgiveness
4. Prayers for salvation
5. Instrumental prayers

Implicit in these routines are images of the divine and a theory of divine action. The following five sections, surveys of these routines, provide a vocabulary and a theoretical background for the study of the religious life cycle of individuals.

I. Generalized Affective Prayers

The first of the five generally common types of prayers is the most generalized and the most affective. It is the prayer which simply praises the Lord, Mary, or other divine figures, or makes professions of love. In such prayers no response is necessarily expected other than, perhaps, a return of generalized love. The proposition repeated every once in a while in sermons or in discussions that "we are on earth to praise God" epitomizes this stance. The phrase "Hail Mary, full of grace," and "Hallowed be Thy name" are examples of this kind of prayer in everyday devotions. Most hymns sung in the church are of this nature, as are the picayos, the songs and dances to the saints that are performed on the feast day.

Sometimes individuals gratuitously bring to the Virgin or some other active patron gifts or propitiations unrelated to any particular request. Such offerings, called *offertas*, can also take the form of pilgrimages or other mortifications. It is a case of free giving. And it is a common and spontaneous expression of affection and devotion to the divine in the valley. The pilgrimages undertaken for this reason are referred to as pilgrimages "por devoción." In this they are distinguished from pilgrimages undertaken "por promesa" and those undertaken "por diversión" or simply "por ir."

Such prayers and acts are considered to add to the generalized store of divine goodwill for humanity and to help balance out the unfortunate acts against God that people commit. The Church reminds the people almost daily that Christ's sacrifice for their sins is something that can never be repaid, but something they should do their best to repay. So that some gratuitous devotion can be seen in the light of a sense of profound debt to the divine. As one woman expressed it, "It is if I owe someone a million pesetas. That is impossible to repay. I just do the best I can. The only thing I worry about is whether I pay as much as I ought to."

Devotion of this type may take the form of conversations—mulling over problems in the church or in the home in front of the image in question. Another form is the "jaculatoria" or "plegaria," a brief, fervent statement of praise delivered spontaneously to the image.

The generalized, affective devotion would be paid to those sacred figures to whom the individual was most particularly attached. They would include any of the figures popular today.

II. Prayers for the Fulfillment of the Annual Round

Gratuitous prayers shade off, in their most instrumental aspect, into another type of prayer common to the valley, the generalized prayer that seeks no specific response, but requests the maintenance of the existing order. Again, as in all types of prayers, this type is used both by individuals and collectives. Examples for individuals would be the prayers said at getting up in the morning, going to bed each night, and when starting out on a trip. Examples for collectives would be the family rosary, the village rosary, parts of the liturgy of the mass, and above all the calendrical village ceremonies to the various divine figures on their feast days. They would include the rogation ceremonies each spring (the greater and the lesser litanies), when the priest leads the village out in procession to bless the fields. These are the prayers which, in a sense, make the year go round. As prayers for guidance and protection they are especially appropriate for the active patrons of the village. They call for the maintenance of village concord and community, the successful unfolding of the environment, and the avoidance of trouble. The prayers request the perpetuation of what is and the fulfillment of what is to be.

Visits on a regular basis to chapels by villages for the annual settling of accounts with the active patron are collective acts and in some ways are much more than the sum of the acts of the individuals. Only this explanation accounts for a certain anxiety for a good turnout and the manifest purpose of the organized processions and dances to the image. But these occasions also serve individuals. Not only are they the occasions to bring the candles and alms that have been promised during the year, but it is a personal settling process, a psychological reassurance, a reconfirmation of a personal contract with the divine. We know this because for many women in villages without active patrons (like Sebastian) the annual visit to the shring is a psychological necessity, regardless of promises and in spite of its being an individual journey. From San Sebastian women go annually to Our Lady of La Luz (June 24, September 8), Our Lady of the Lowlands (August 15), Our Lady of La Salud (September 8), and even to Our Lady of El Brezo (September 21). Different women go to different shrines. But possibly a third of the adult women, say 20 to 30, go to one or another shrine, the same shrine, every year. Listen to Gerónima of San Sebastian describe her relationship with the shrine of La Salud in Puentenansa, two hours from her village.

> I have been to La Salud every year, except for a couple.
>
> How old were you when you first went to La Salud?
>
> I first went when I was younger than [her daughter age 12], with my mother. My mother always went. I have never gone there because of a promise. One time the day fell on a Sunday, so I went to mass here, and coming out of mass I felt a sinking feeling [points to chest] so strong that I had to go. So I took off straightway and arrived there and said three Hail Marys and came back. Another time there were two people in the shop, and so I didn't go. And I felt terrible, and I said to myself. "Are you not going just in order to make money?" I got all upset and tense, and I had to go. When I go I take my children, just as my mother would take me.

As Gerónima describes it, her relationship to La Salud is so strong that it has physical manifestations. It involves not only a loving relationship with the Virgin herself, but also a compulsion to make the trip to the shrine. But her compulsion to do it regularly, every year, on a certain day, would seem to be her way of fulfilling a regular contract, making a calendrical contract with the Virgin, just as the villagers of Tudanca go to the shrine of Vado de la Reina every year on August 5. San Sebastian, unlike Tudanca, has no chapel outside the village and no active patron. Gerónima is an extreme case, but there are others like her. At the present time few men (if any) from San Sebastian are so regular in their devotions, and I do not know about them in the past. But clearly this kind of compulsion is operative in the collective trips. Note that Gerónima has no formulated reason. She goes, in her terms, because she has to. It is our inference that what is involved is a unspoken contract, that somehow the nature of her relationship to the divine and the divine attitude towards her partly hinges on her fulfillment of her customary visit. Somehow for her the visit has become part of the very order of things.

Jaime, of Tudanca, an otherwise desultory churchgoer who says he believes in little or nothing, shakes his head when he remembers that some years he did not visit the chapel on August 5. "The worst thing I ever did was not to go to the shrine on August 5." He was very serious, feeling that he had failed to meet his obligations to the Virgin. For him that was the one thing in the religion important enough to hold on to, a sacred contract. It was the only thing in the whole gamut of devotions we had discussed that he had considered important or relevant.

Time and again the importance of the obligation owed collectively and individually by the village and villagers to their active patron on that patron's fiesta was impressed upon me. The responsibility that the villagers felt to their patron (usually Mary) seemed to be born more out of respect than of fear. It was as though the Virgin would be hurt if they did not turn up. They were doing it for her as much as for themselves. It is this sentiment which demonstrated to me the depth and the personalness of many people's feeling for the Virgin, in spite of the scepticism of many of the village priests on the subject. The phrases used to describe their relation to Mary (Our Lady) were not unlike the phrases used to describe their relation (in one village) to be a beloved and respected nobleman. "Lets pay a call on the Virgin," is something one frequently heard.

These calendrical devotions, then, involve an amalgam of affection and obligation. The sense of obligation, of course, has been nourished for centuries by the Church and at times enforced by the civil authorities. At present it is still strongly held by many in the valley, although much less so than at the turn of the century.

III. Requests for Forgiveness

A third type of prayer, the request for forgiveness from wrongdoing, seeks a definite response, unlike the first two. The response is, however, unverifiable. In its most generalized, collective form, the prayer shades off into the kind of all-purpose request for protection discussed above. An example of this would be in the rosary, "Pray for us, sinners." A more specific form of collective penance would be a procession against a scourge of one sort or another that was dimly understood to be a punishment for the wrongs of the community. Another would be a communal act of atonement, undertaken by priest and parish in repentance for desecration or defilement of the holy place.

In its individual form, involving confession and communion, it is a sacrament of the Church. The discussion of the life cycle (below) reveals the frequency of such prayers and the person likely to use them. This kind of prayer can also take the form of a penitential trip to a chapel, but this usage appears to be very infrequent. The ancient church assigned pilgrimages as penances, and they were still common in the middle ages, but this is no longer the case.

IV. Prayers for Salvation

A fourth type of prayer is a request for divine action in the world to come. Prayers for forgiveness naturally shade into this kind of prayer. In its most general, collective form, this could include the Our Father that is sometimes said for the souls in purgatory at nightfall when the handbell is rung through the alleyways and contributions for masses for the souls in purgatory. In more specific forms it would involve masses and prayers said for the dead in the family, prayers and acts calculated to reduce one's own time in purgatory and the application of indulgences one has obtained to specific souls in purgatory.

Indeed, the whole institution of indulgences, whereby persons performing sincerely certain devotional acts are awarded commensurate temporal pardons, serves to limit the sense in which almost any devotional act can be a purely gratuitous one. Virtually all of the generalized devotions and most of the regional shrines figure in the system of indulgences and earn their devotees fixed amounts of days of indulgences for each visit or each devotional act performed. There is ample evidence that the concession of indulgences by papal bull or by other ecclesiastical authorities has had an enormous effect on the growth of the devotions in question. The shrine handbooks, the rules of the brotherhoods, and the devotional handbooks all emphasize the indulgences earned by their respective devotional acts.

While it is clear that these "bonuses" played an important role in the past, at present their importance is very slight. In all my conversations with valley residents, never did the topic of indulgences arise. In the past 15 to 20 years it seems to have dropped out of importance as a religious motive. No doubt this is partly due to a dramatic downplaying of the entire notion by the Church. The use of "days" as a measure of forgiveness for simple acts was a relic of the former penitential discipline of the Church whose meaning for the layman was never precisely defined. What the people understood (and this mistakenly) was that these were days earned as time off from purgatory and that plenary indulgences would mean no time at all in purgatory. Although it is not a topic of interest now, surely such a system must have its effect on present-day concepts. The very notion that in performing acts of devotion one is building up one's store of treasure with heaven must be the basis for much of the gratuitous devotion. And surely the practive of granting indulgences has played a large part in supporting the notion that devotions, sins, and acts are translatable into numbers, into some absolute-value set of equivalencies. This notion lies behind much of the scheme of reciprocity and interrelations with the divine, especially in the fifth and final category of prayers.

V. Instrumental Prayers

The most instrumental and most specific kind of communication with the divine seeks a response from the divine in the form of action in this world about a particular problem. These are petitions of people to God in times of crisis. They are, by the nature of the situations that engender them, the most dramatic form of prayer and the form of prayer in which the valley as a whole has the most interest. The most common crises involve sickness and accidents, especially of children; the straying or the sickness of livestock; the potential injury of a son in military service; and other personal and family crises. For small problems and normal sicknesses, the normal intermediaries to whom these prayers are directed would be the souls in purgatory or the active patrons. For problems of livestock the intermediary would almost exclusively be Saint Anthony. For other grave crises the intermediaries turned to are always the active images of the shrines. The overwhelmingly important mediator at the shrines is, of course, Mary.

In the upper Nansa valley this type of instrumental prayer takes the following forms:

A. The "Thy will be done" variety. This prayer engenders no obligation on the part of the person making it, and by its very nature it is self-fulfilling. Psychologically it is perhaps the hardest of prayers to make. The dynamic behind this type of prayer is that the most efficacious position to take in regard to the divine is that of faith and trust. That was what Jesus demanded time and time again before he would perform the miracles requested of him. Incidentally, this underlying principle that faith is a necessary component of effective communication with the divine plays a role also in the calendrical devotions, in which the trip to the shrine itself can be a statement of faith.

B. The specific request, without a concomitant pledge (e.g., please cure my son). If the request is granted any of a number of options might be taken in repayment, called *acciones de gracias.* On the village scale this is the kind of prayer made when the village is assaulted by drought, wind, plague, or war. Such corporate requests, called *rogativas*, were usually made as processions to chapels. They were common in this valley up to the twentieth century but are never seen now. They are still used in other parts of Spain, especially for rain on the meseta.

C. The most common form of petition prayer is known as *promesa* (promise, pledge; in Latin, *voto*). This is a conditional pledge that specifies what reciprocal action the pledger will take in the event of a favorable outcome. "If you cure my son, I will go to Covadonga barefoot." This is the prototypical prayer of the valley and of Mediterranean Roman Catholicism. It is the prayer that is behind the establishment and use of many of the chapels, and it is the prayer that best defines many persons' relation to the divine.

The principle behind the promise seems to be related to that involved in the sacrament of penance. Penance assigned by the priest after confession involves the sinner giving something up to redeem his sin. Now the penances assigned are usually symbolic, in the form of prayer, but in previous times they often involved pilgrimages or other mortifications. Promises involve payments in goods or effort to the divine in return for out of the ordinary divine attention.

Both involve a restoration of a natural balance. Penance is out of the ordinary human action that equilibrates a system that the individual himself has thrown out of order; the promise balances accounts with the divine for voluntary action on the part of the divine which had disequilibrated the divine-human relation.

The promise has many variations. It generally involves a pledge of some sacrifice of resources by the promiser. This may be expense of money in the form of masses, candles, or alms; expense of time in prayer, novenas, for instance; expense of time and energy in a pilgrimage; or the sacrifice of pride or conformity—cutting off hair as a gift or the wearing of a plain sacrificial habit. It may also involve the denial of pleasures, such as dancing, or perhaps the public testimony of the act of God that occasioned the fulfillment of the pledge in the form of the commissioning of a painting, the lighting of candles at a shrine, or the testimony of the devotee in a devotional magazine. Promises may be made by proxy for the person in danger to fulfill, and they must be fulfilled, whether by the person who made them or by someone else.

In terms of money, the most commonly made pledges are given at the alms boxes in the parish churches. These alms boxes are overwhelmingly used for pledge money, money given in return for a positive divine response to prayer. The most common alms boxes in the valley are those to the souls in purgatory, Saint Anthony, and the Virgin of Fátima. The magazine most commonly used for pledges is *El Santo*, the magazine of Saint Anthony of Padua. Every issue lists money sent from this valley. Masses most commonly offered to divine intermediaries vary considerable from village to village; most generally they are to the active patron. And the use of money pledges is most common at the shrines. At El Brezo, for instance, the traveling throne of the Virgin is specially designed to catch money as she travels around the shrine after the mass. At La Luz everyone files up to kiss the medal hanging from the statue and leave a coin in the plate. And the other shrines have similar provisions for the contributions of pledge money from the people.

Virtually everyone in the valley with whom I talked had left money from pledges. Many did it frequently. The use of novenas is harder to evaluate, because these are private devotions performed by individuals alone, generally in the parish church. At the major shrines it used to be customary for persons to stay nine days for votive novenas, but except in a few places in Spain this is no longer a custom, certainly not around this valley. As we said above, novenas are more particularly used for petitionary devotion, for the premature payment of a hoped-for favor.

Promises to make pilgrimages are made predominantly, but not exclusively, by women, especially mothers. They may include additional stipulations, that the promiser go on foot, or barefoot, or around the church on her knees, or without speaking, or without eating or drinking. Usually, nowadays, it simply involves the trip, possibly by bus if the shrine is far. There are cases of promises to visit a number of shrines. The promise can include leaving an offering, or *ex-voto*, at the shrine. The custom dates back at least to pre-Christian Greek religion. In this valley the gift is almost exclusively money or candles, but in other parts of Spain the gifts include wax models of the parts of the body cured; hair; photographs; paintings of the grace received (accidents and sicknesses especially); and even the fruits of the land, in the form of corn, animals, or other goods. In the Obeso death records many wills contained offerings of animals, land, and other goods to the image at the shrine, and clearly some of these offerings, like some of the death masses, were debts the individual had built up to the divine over his or her lifetime, much as Socrates had a debt of a cock to Asclepius outstanding when he died.

The shrines listed in Tables 8 and 9 are the most common destinations of pilgrimages from San Sebastian and Tudanca:

TABLE 8

Favorite Shrines of San Sebastian People[53]

Shrine	Distance (km)	Women (11) (trips)	Men (6) (trips)
Our Lady of Lowlands	$5\frac{1}{4}$	49[a]	3[a]
Our Lady of Luz (Peña Sagra)	$12\frac{1}{2}$	26[a]	40[a,b]
Our Lady of Good Health	$7\frac{1}{2}$	24[a]	6
Our Lady of Covadonga	95	13	2
St. Toribio (Holy Cross)	53	9	0
Our Lady of Brezo	85	5	1
Christ of Bielba	$18\frac{1}{2}$	5	1

[a]Some people in sample could not accurately remember the number of visits to these nearby shrines. These numbers are conservative estimates.

[b]The men went especially to La Luz shrine because they tended cows on Peña Sagra, where the shrine is located.

[53]These tables are based on interviews with people of all ages in San Sebastian and Tudanca. I asked them about each visit they had made to each shrine in the region in their lifetime, trying to get them to recall about how old they were (or what year it was) and whether or not it was for a promise. Thus the visits to shrines recorded in these tables took place in the entire timespan of living memory (approximately 1900 to the present). When the trips are arranged by year, rather then age, the rate of trips and the destinations appear to be fairly constant throughout the century, with two exceptions: One was the brief hiatus of the civil war, when the trips virtually ceased; the other is the period from 1956 to the present, which shows a marked increase in visits to shrines. Part of this increase might be explained by the fact that people have a better memory of more recent trips. But there are other factors at work. One is an increase in parish excursions, with chartered buses (this to

TABLE 9

Favorite Shrines of Tudanca People[a]

Shrine	Distance (km)	Women (13) (trips)	Men (30) (trips)
Our Lady of Vado de la Reina	$3\frac{3}{4}$	Innumerable	Innumerable
Our Lady of Brezo	72	16	13
St. Toribio (Holy Cross)	54	16	5
Our Lady of Luz (Pejanda)	9	11	1
Our Lady of Covadonga	103	9	4
Our Lady of Luz (Peña Sagra)	$16\frac{1}{2}$	5	1
Our Lady of Bien Aparecida	133	5	0

[a]The Tudanca people, with their own shrine, pay far less attention to shrines in the valley than San Sebastian.

The villagers have problems of greater or lesser magnitude. Herein lies the secret of the flourishing of so many shrines. Many villagers have made promises to different shrines, the more important promises made to the more distant shrines. The principle involved is that the effort expended to reach the shrine must be commensurate with the favor requested. For instance, a lady in Obeso with a slight worry might offer to have a mass said at the local shrine of Our Lady of the Lowlands on condition the worry was solved; a successful mending of a broken limb might be worth the promise of a trip to Bielba on the fiesta day; and the release of her husband from prison might be worth a promise to El Brezo, three days walk away. For this reason, few promises are made to the local shrine. The local active patron is attended to for the successful prosecution of the normal annual round; only if getting to it is a special effort, as in the case of Las Lindes at Carmona, Vado de la Reina at Tudanca, or La Luz on Peña Sagra, all a considerable distance from their villages, will the villagers make promises to the image that involve physical effort. Generally the greater the promise, the farther away the shrine. But if time cannot be afforded for a long trip, closer shrines may be used for more important promises by adding obstacles to the trip—promising to go barefoot, without speaking, without drinking; or by the addition of other penances such as the purchase of masses or exvotos or the saying of novenas.

---

the more distant shrines); another is increased prosperity, which means more leisure time for trips, and possibly more inclination.

A technical note: The tables of trips by age have fewer trips listed because they do not include annual trips to the shrines of Salud or La Luz of two persons interviewed in San Sebastian. Inclusion of these trips would have distorted the tables.

Evidence from the history of Obeso shrine devotion indicates a three-tiered system of shrine devotion: The inner tier is comprised of the shrines in the village (Our Lady of the Lowlands, especially); the second tier is comprised of shrines in villages close by to the south (La Peña, and the Christ of Bielba); and the third tier is comprised of shrines one day or more distant in the mountains (like El Brezo). Evidence that such tiers existed can be seen in the graphs of shrine devotion over time, which show shrines of equal distance becoming popular when older shrines begin to wane. Similarly today promises are offered from San Sebastian either to the nearby shrines of Lowlands and Luz or to the regional shrine of El Brezo. Very minor problems are taken to the souls in purgatory in the Church. The two outer tiers might well apply to two discrete levels of problems, perhaps something like mortal danger for the outer tier and nonmortal danger for the inner tier. The local tier might be of the more petty, day-to-day worries.

We know, therefore, that while certain persons stick to one personal patron above all, usually one particular Mary, others divide their devotion on a somewhat utilitarian basis among images at different distances. The first solve problems of different magnitude at the same shrine by varying the cost of the thank offering or the difficulty of the trip that they impose upon themselves. The second simply apply problems of increasing magnitude to increasingly distant shrines. Many persons use both systems, and still others have very little to do with shrines at all.

Table 10 (see also Figs. 9 and 10) shows the age and sex groups of those most frequently making trips to shrines. The difference between women and men in devotion to shrines shows up clearly. Both men and women, during the period of courtship, go to shrines, perhaps primarily for the fiesta. But after marriage the visits of men to shrines drop off to virtually nil, while those of women continue at a high rate through the period of fertility and child raising, dropping off gradually as old age renders them less and less able to get out. Visits of women to shrines after marriage are primarily for the welfare of their children. These visits, made in the age range 30-45, are the ones most likely to have been made as the result of a promise.

The second manifest contrast is that between customs of San Sebastian and Tudanca. Tudanca has its own shrine, and everyone physically capable probably visits it several times each year. The Tudanca people therefore have less necessity than the San Sebastian people to travel to outside shrines.

Table 11 breaks down the destinations of trips to shrines by the distance of the shrine from the home village, and the age at which the village woman makes her trip. From both villages, women's visits to regional shrines continue at a steady rate until women are in their late fifties. These are largely for important promises. For women under 35, trips to nearby shrines may be for diversion (courtship, dancing) as well as for devotion. Hence there are more trips in these

TABLE 10

Reported Visits to Shrines by Age, Sex, and Village

Age at trip	A: No. of persons surveyed who were in or had been in age range				B: Total trips by persons at these ages				B/A: Trips per person during age period			
	F	M	San Sebastian	Tudanca	F	M	San Sebastian	Tudanca	F	M	San Sebastian	Tudanca
0-14	23	13	17	19	12	1	8	5	0.5	.08	0.5	0.3
15-24	22	13	17	18	39	23	38	24	1.8	1.7	2.2	1.3
25-34	17	12	16	13	21	15	38	8	1.8	1.3	2.4	0.6
35-44	17	12	16	13	23	–	18	5	1.4	–	1.1	0.4
45-54	12	12	14	10	17	1	10	8	1.4	.08	0.7	0.8
>55	7	6	8	5	12	–	3	9	1.7	–	0.4	1.8

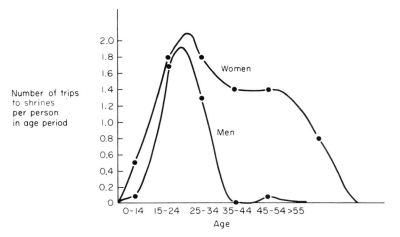

Fig. 9. Graph of visitors to shrines, by sex.

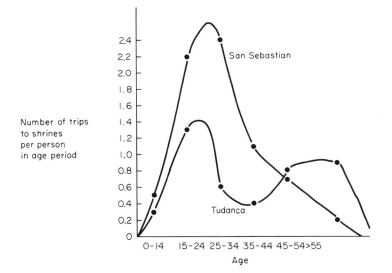

Fig. 10. Graph of visitors to shrines, by village.

TABLE 11

Destinations of the Pilgrimages of Women by Age at Trip

	Vale and valley shrines (5-20 km)	Regional shrines (50-133 km)
San Sebastian (N = 11)		
Under 35	36+	10
35 and over	20	11
Tudanca (N = 13)		
Under 35	13	14
35 and over	1	18

years. After marriage and the age of courtship, trips to nearby shrines are primarily for devotion, especially for smaller promises. Tables 10 and 11, like those before, reflect the contrast between villages with their own shrines and those without them. Tudanca's local shrine (not included in the figures) takes care of the smaller promises and the visits for devotion. Hence the only visits from Tudanca to valley shrines are for diversion. This explains why the Tudanca women make relatively few visits to valley shrines during the *mocedad* and virtually no visits to the valley shrines after the courtship period. After age 35 the only nearby shrine visited by the Tudanca women is their own. San Sebastian, in contrast, is one of the villages that has no shrine. Hence devotional visits that the San Sebastian women would have made to their own shrine, both during and subsequent to the years of courtship, must be made to the shrines in other villages.

Visits of men to shrines, as we have seen, take place primarily during the time of courtship: they are limited for this purpose to the closer shrines, those in and around the valley.

The wearing of a penitential habit is an alternative kind of promise. The habits listed in the following tabulation are most popular in the valley. These habits must be of pure wool and may be worn publicly or under other clothes. Some women wear the habits only on fiesta days; others have made promises to wear them all the time for a certain length of time. The wearing of a habit generally entails a series of supplementary obligations (no dancing, no participation in profane entertainment, etc.), all of which make it evident that the wearing of a habit signifies a protoreligious state, a mimicry of monasticism. It will be noted that the devotions to which the habits refer are generalized devotions, devotions to images to whom pilgrimages cannot be made. Hence the wearing of a habit seems to be the devotional equivalent of making a votive pilgrimage to a shrine. But the monastic aspect of the practice is instructive,

Habit	Color	Popularity (roughly)	
		Degree	With whom
Our Lady of Mt. Carmel	Brown	Most popular	Women
Our Lady of the Sorrows	Black	Most popular	Women
Our Lady of Perpetual Help	Purple	Moderately popular	Women
Sacred Heart of Jesus	White	Less frequent	Women
The Immaculate Conception	Blue	Less frequent	Children
St. Rita	Black	Less frequent	Women
St. Anthony	Grey	Less frequent	Men and women
Jesus of Nazareth	Purple	Less frequent	Men and women

because it shows just how deeply the monastic tradition of simplicity and sacrifice has colored all of these religious devotions. I do not know just how active the religious orders were in cultivating the wearing of habits. I find no reference to habits in devotional literature (aside from a mention of indulgences for wearing the cord of St. Francis around the waist) and only scant mention in the parish records (the traditional custom of wealthy persons being buried in the Franciscan habit, in Polaciones). So I have no idea how old the custom is. I am told by the clothing merchants that the custom is dying in all but the remote villages. At any one time in San Sebastian there might be four or five women wearing habits, although some of them wear them under their other clothes. In Tudanca it is also a custom, but one that is seen less frequently. I know little about the occasions for the promises which led to the wearing of these habits, for that too is private information which my friends did not volunteer.

How are promises, personal contracts with God, enforced? Local and national popular religious folklore, doubtless encouraged by a long line of shrine tenders, has it that dire consequences follow from the nonfulfillment of promises. In each book of miracles published at the different major shrines from the sixteenth century on and in the devotional literature distributed in favor of one saint or another, there is usually at least one cautionary tale of some poor fellow who forgot to fulfill his promise. In fact the example is almost always of a man, as men seem to be more generally lax in these matters, making few promises to begin with and tending to postpone their fulfillment.

My landlady and friend in San Sebastian told me the following tale. There was a man who had a son who was very ill. After the doctors had given up hope, the man promised to receive communion if the son got well. The son did recover, and the man went into a church. But he had second thoughts, and finally the demons chased him out of the church and he went to a cafe. At that point some friends went by with his son across the street and the man signaled to them. The boy started to run across to him and was killed by a car. And the father cried out, "It is I who have killed him!" The stories in the local traditions surrounding the major regional shrines in Spain are slightly less gruesome. They usually involve only a relapse of the sickness cured, or a recurrence of an accident, at which point the promiser remembers that he has neglected to pay his debt and renews his promise, generally augmenting it.

Perhaps because of the perils of unfulfilled promises, those promises made in the valley are usually couched in a form that does not commit the promiser to fulfill the promise in a certain time span. The promises can then be fulfilled at one's leisure, in a year that is convenient or when the promised donation can be afforded. Some of the more distant shrines are a two- or three-day walk. Such a trip is more pleasant and safer with company. So people save up their trips until they can find a companion or a group to go with. As a result, many of the women in the valley have several promises pending at any given time, sometimes to different shrines. If these promises are unfulfilled in their lifetime they must be fulfilled after their death by the children or friends.

The same perils facing nonfulfillment of private promises face the reneging of collective promises. Such promises are a common feature of western European Catholicism. I know of no collective promises in the history of the valley. But the principle is just the same as that for private promises: a community promises if spared an imminent plague or drought, or locusts, to make an annual procession to the shrine on the day of the dispensation. Or the pledge is made after the dispensation, as a thank offering. But in any case many shrines have traditions that the particular scourge in question recurred as soon as the procession lapsed.

D. The petitionary devotion. Because of man's intrinsic debt to the divine (if for nothing else for the mere opportunity of redemption) it is not considered that the gratuitous acts of devotion described at the other end of the prayer spectrum obligate any response from the divine. And the very way that promises are phrased shows a sensible appreciation on the part of the villagers for divine unpredictability in the granting of petitions. But there are cases in the prayer books of language being used that would imply that supplicatory acts, or acts of devotion, especially for the souls in purgatory, earn credit with the divine for divine actions on this earth. Remember for instance the extraordinary phrase used at the beginning of the novena to the souls in purgatory, "it is a very efficacious means of *putting God in one's debt* to make this spiritual alms . . ." (my italics). And there is a natural tendency on the part of the people to consider that their devotional acts may be useful in swaying the outcome of problems under consideration by the divine.

So a final variation of the specific request is the initial commitment of resources to the divine in the hope that the divine will respond favorably. While there are cases of pilgrimages of petition, especially of the chronically ill or deformed, the most frequent vehicle for these efforts are the novenas, or nine-day devotional cycles of prayers available in the prayer books to various saints. Another way to do it is to have a mass said "for the intentions of an individual." The custom of petitionary masses varies from village to village, common in some, infrequent in others, depending, it seems, on the degree to which it is countenanced by the parish priest.

These prayers have been arranged analytically for later comparison with the modes of exchange among humans in the valley. But they seem to represent the mixture of two systems for regarding the divine. The instrumental prayers and the prayers for the fulfillment of the annual round (types 5 and 2, respectively), both of which call for divine action in this world, belong in an old tradition emphasizing immanence, essentially based upon the shrines. The prayers for forgiveness and salvation (types 3 and 4, respectively) are moral prayers, usually made through the intermediary of priests. These prayers involving divine intermediaries are more likely to be the generalized devotions. My guess is that this kind of prayer came into the valley at a later date than the instrumental prayers. The generalized affective prayers (type 1) are found in both systems of regarding the divine.

The first of the two systems sees the divine and the human as two separate categories that can treat with each other under certain conditions of precedence. The second system regards the human enterprise in its most exalted state as progress toward divinity, the process of achieving some kind of holiness. For the former the ideal behavior is the dignified fulfillment of the amenities, always maintaining a kind of separation, as with interactions between humans. For the latter system the ideal behavior is for humans to act like the divine as much as possible, indeed eventually to become divine.

Until the advent of the young priests of the Second Vatican Council (whose impact is studied at the conclusion of this section), these two systems, essentially different, have coexisted and interpenetrated in the Roman Catholicism of the valley. Most persons in their religious practices have partaken of both systems, but some have adhered more to one than to the other mode of approach to the divine. This has happened in such a way that in the different age and sex groups of today the successive world views of the distant past are available like so many geological strata. For the system of purificative, salvationary religion has "taken" more with some elements of the population than with others.

Individuals will differ in the extent to which they offer the types of prayers and the various petitions. In addition, each individual is born into the valley with an idiosyncratic accretion of traditions of worship—from his family with its own selection of divine patrons and its own collection of pictures, literature, and prayers; to the preferences of the priests, missionaries, and teachers who help educate him in religious matters; to the village with its own distinctive devotions; and to the devotions of the region, the nation, and western Christendom in the years he or she is young, as brought in by magazines and other media, missions, or returning villagers. For these reasons the "active pantheon" of each valley person differs: which saints are revered; which devotions are performed; and the priorities that obtain among them. Finally, individuals will radically differ in the salience of religion as a whole in their lives. Some of the villagers do not have interest in any of the images and devotions, while others devote virtually every spare waking hour to some form of worship. Most fall somewhere in between. On this point—the place of religion in the life space, as it were—every person again cannot fail to be distinctive.

Yet while every person's attitude toward religion is inevitably idiosyncratic, it is possible to make rough generalizations about groups of people according to their different conditions and to note critical stages in the life cycle when contacts with the divine are most likely to take place. Such regularities come clear from conversations with the villagers and participation in their devotional ceremonies. I came to know about 15 persons in the valley in such a way that I felt free to ask them about, and they felt free to talk about, their contacts with the divine, or lack of them; their beliefs; and their devotional history. From these persons I could also learn whether or not they thought they were typical of persons of their sex, age, and condition. Of a larger number of people in Tudanca and San Sebastian I asked what shrines they had visited, at what age, and under what circumstances. Some of that information has been presented in tables above. And of course I could see who went to mass and rosary, who received communion, and who went to the shrines on the ceremonial days in the years I was there.

From these experiences and observations it is clear that there are certain more or less regular accommodations reached with the divine authorities that seem to vary with age set, sex, isolation from the community, and relation to the world beyond the valley. To mitigate somewhat the dead hand of sociology that turns interesting individuals into uninteresting averages I will also report in detail the experiences of some of those friends who fall into the different groups.

Many adults, perhaps a majority in the valley, hold a special relationship to one sacred figure that they feel particularly close to. There seems to be no commonly used term for this relationship. For future reference I will call the divine figure specially revered a *personal patron*. In villages with strong active village patrons like Tudanca, Obeso, and Carmona, the active village patron is generally the personal patron. As such the personal patron is learned in early childhood and carried through to the grave. In villages with less dominant active patrons (for instance, San Sebastian or Rozadío) the personal patron is not likely to be the village patron and is not chosen until it is needed. In both kinds of village the points in the life cycle at which the personal patron is turned to, and religion itself is taken seriously, are the same.

As the Rionansa child comes to awareness, he or she learns that there exist beside the family and other people additional beings which, while not visible in the flesh, are represented by images on the church altars and colored pictures in the bedrooms and kitchen of the house. These pictures differ from the photographs of the parents at their wedding. The photographs of the parents are solemn, formal pictures, almost lugubrious with dignity. The pictures of the other beings are colored, bright, and pretty—a lady with wings helping two children to play—and the faces show expressions—a man in a robe straining under a cross; a beautiful lady looking up at the sky in wonder; a gentleman holding a child in his arms.

From an early age, then, the idea is implanted that the divine figures are benevolent and sympathetic. But when children are younger, although they are exposed to the idea that God can help them in a variety of ways and although some of them have more of a sense of nearness to God than they do later as adults, they do not need a personal patron. They know that their mother has a favorite saint that she prays to since this is usually no secret, but they themselves have not felt the need, nor are they particularly encouraged to. My conversations and the letter columns of devotional journals dedicated to specific saints indicate that it is not until mid or late adolescence, if then, that the child focusses prayers on a personal patron. Before that, one has not been needed. The parents, especially the mother, have handled the child's problems, have served as protectors.

In their teens, people begin to have transcendent problems that are all their own, that they must guard even from their parents, like finding, keeping, or retrieving a boyfriend or girlfriend. This is most especially the first time that patrons are used and the time that in the villages without obvious active patrons, teenagers choose patrons. One set of persons who are marked as especially precocious in this matter are those who have been chronically ill in childhood or adolescence. They seem to find patrons sooner, needing them sooner. On the other hand, those who have a smooth courtship may not have to turn to the divine for special assistance until some later catastrophe or danger befalls them. But courtships are rarely smooth. Few are those who are not, at one point or another, disappointed, for into the round of courtship, dances, fiestas, and the mocedad goes the major part of a person's interests, hopes, and dreams from about age 15 for girls and age 17 for boys until marriage. As the parish priest of Celis explained it, "A more vital faith, a real encounter [with God] usually begins when the normal sequence of life is broken. When something goes wrong, then the person has to stop and think. People accept only the truth that is convenient. They must change before the truth is accepted. The truth itself does not change. For some people this never happens. And as it varies in each particular case, it is difficult to fix a time to. Something like this is called conversion in other religions. It might come because of a sickness, or a death, or a girlfriend." He is a modern priest, who emphasizes a kind of devotion more modern than that involving personal patrons (Catholic Action, Cursillos de Cristiandad). But his description of the conditions for conversion perfectly fit the occasions of choices of personal patrons that people in the valley have related to me.

The missions are a case in point. They have as one of their purposes a reawakening of the parish, a break in the normal sequence that will produce as one of its consequences a certain number of these conversions. And the particular devotion that the missionaries bring along is the new personal patron to whom those whose devotion they have quickened will turn. It seems clear from the literature distributed by the missionaries that their emphasis upon the more unpleasant possibilities after death is calculated to jolt the villagers into a new awareness, leading to a new level of consultation with divine intermediaries. The principle behind conversions by way of missions and in the normal course of the life cycle is the same. A sure guiding hand is sought at moments when the individual is most disoriented and at the same time most alone.

That a special patron is not turned to until one is in a jam or under pressure does not imply that a person has not been devout or that patrons are looked to merely for protection. It is simply a trigger that releases the habit of turning to outside help at times when a personal, often private, need is felt. Those of us who have been close to others know that we come to depend on them, sometimes developing a need to fit a relationship, instead of *vice versa*. When we know that there is someone around to help, we call for help, even though we might perfectly well get on without it.

The metaphors of parenthood for the Virgin and Christ are surely instructive. It is certain that a personal relationship that develops between the person and the divine most closely resembles that between parent and child, and it often develops, as we have described it, at a stage when the individual is moving out of the shelter of parental protection. Mary and the other more universal saints, such as Saint Anthony of Padua or Saint Francis, have undergone a somewhat revolutionary humanization of their image since the middle ages and are accessible not merely as sources of power to be tapped, like the Lord of the Manor, but also as parents or uncles, to be loved and revered. It should not be forgotten that all these divine figures were once human, and it is on the basis of their humanity that they can be approached as intercessors with God. Therefore the search for counterparts in human relations to the relations with the divine starts from the reasonable premise that the divine are seen as human. They are addressed as human, and when they appear, they appear as humans, powerful and good persons who are attentive to the needs and well-being of those on earth. There is not, then, from the perspective of the valley, a radical separation of the divine from the human. Rather there is a long, helping chain that goes from the sinner to God, by way of the souls in purgatory, the saints, and Mary.

The habit of fixing on a personal patron among the local pantheon of saints would seem to be a very common feature of rural Catholicism, yet it is rarely discussed either by anthropologists or by the Church. It is not necessarily true that only one patron is revered, but generally there is one preferred. Some people well remember the moment their own choice was made. I spoke about this with a woman born in San Sebastian, now living in Obeso, and she described her own case:

> I have always been specially devoted to the Virgin of Brezo. I can tell you how it began. I remember I was walking with Mariana, the woman in San Sebastian who has gone to El Brezo so many times. I remember precisely that we were at the place where the big rock juts out on the path before crossing the bridge near San Sebastian. It was about nine years ago [when she was 31]. Mariana said, "Going there moves me so much!!" and I had a kind of jump in my heart and I said, "If I am cured [she had a case of phlebitis] I will go to Brezo." It's funny because my mother was always talking about Brezo, and it never made any impression on me. And I even went there in a bus on excursion when I was younger, and it made no impression. But since that time I have always had a great devotion to the Virgin of Brezo.

In another case of a San Sebastian woman (recorded below) the Sacred Heart of Jesus appeared in a dream every night for a week after her husband died, declaring his protection. And in many other cases from other villages the choice was made in early childhood by the mere fact of being in a village with a strong active patron. The devotion in this kind of village seems to be passed on from parent to child, especially from mother to daughter.

Women are more likely to fix on personal patrons than are men. This is true especially with regard to her children—at childbirth, when they are ill, or when they go away from the village. The woman has more to ask of her patron after than before marriage. From the simple *jaculatoria* "Aie! Madre de las Nieves!" when she is frightened to the novena in the church, or the wearing of a penitential habit, or the promise of a visit to the shrine, she has available to her a great number of ways to call upon her celestial aide. The overwhelming proportion of promises are made by women of child-bearing age. Such a close relationship carries over into old age when practical requests from the women still center on the children, now grown up, and when there is a deepening of the personal, affective ties between the women and the divine patron.

Although some of the men, also, have personal patrons, virtually all transactions with the patrons are handled by the wife. This seems to be part of the village and family division of labor, equivalent to the apportioning of other tasks to women, children, or men. A couple of herdsmen explained the theory to me in a winter barn:

> Women make promises. Do men also make promises?

Yes.

Does the man or the woman usually do it?

The woman. Usually the woman says she is going to do it and the man agrees to it. For instance the wife might declare, "I'm going to El Brezo with the boy," or with the husband, if the promise was made for him.

Why the woman?

It's like when in the kitchen the woman says, "Let's recite the rosary," with the husband and children there, and so we say, "All right, let's recite the rosary." In these matters it is the woman who initiates things. It's like when I go and tell the wife and children it is time to sell the cows. There isn't a reason. There is a reason to sell them, but it is the man who decides to do it. The woman might say to the man, "Let's have a mass said at the shrine."

The woman, then, assumes control of all affairs pertaining to the spiritual well-being of the household: the masses for the dead, the children's prayers, the husband's annual communion, and the negotiations with the important divine figures. Just as the man is the family head in practical matters, the woman has a certain responsibility and authority in spiritual matters. For instance, in matters involving the health and safety of all family members, the wife or mother will be the one to make the promises, whether to her own personal patron or to an appropriately distant shrine. Examples of such proxy promises that I know of include times when a son goes to war or to peacetime military service, when a husband is jailed, a child is sick, or a family member is in an accident. Some cases will be given in the course of testimony later in this chapter. For several men the only time they had been to a shrine was when their mothers had made a promise for them to go if a crisis worked out satisfactorily. One older man had only been to El Brezo once, and that was because his mother, a devotee of El Brezo, had promised for him to go if he got back safely from the Moroccan Wars. He had neglected to write her often enough, so she became worried.

With the promises made for them by their wives, regular trips to the local shrine on the feast day, or trips to shrines as social occasions as young men, the men too participate in the devotion to and the maintenance of the shrines. But the division of labor virtually obviates the village man's need for a personal patron. Just as previously the major crises were handled by his parents, now the major crises are handled by his wife.

One class of men that stands out in exception is that of priests. All professionals maintain devotion longer, partly because of an education which in its later years was almost exclusively Church-supervised. They receive communion more frequently, they are likely to sit in the front row of benches in the male section, and they have a sense of obligation to and an affection for the village patron. This is especially true, of course, for priests, for whom the village Mary has generally been an aid in finding their call. The establishment of a call, or vocation, is an emphatically personal enterprise, and as such a matter for the boy and God. In his many years of seminary it may be questioned time and again. And we have seen that strong relations with the patron develop precisely when people are most alone with their decisions and their questions. Priests, in a sense, have no home. They certainly have no wives to take over their negotiations with the divine. Nor do they have the sensitivity wrung out of them and the hardness instilled in them that normally happens in the course of attaining manhood in the village. They are better able to preserve affection for Mary, and in seminary they feel no need to be ashamed of sentimentality.

I do not know whether persons change their personal patrons. I am inclined to think that cultures change patrons but individuals do not, that is, that patrons decline through attrition, not abandonment. The graph lines of the Obeso devotions support this idea, as they drop off gradually while new lines are rising, indicating that persons hold on to a devotion until they die. It is hardly likely that fads would sway the intense relationship that grows between a person and his or her personal patron. Yet fads do operate on a higher level, being responsible in part for the presence of certain devotions within the range of options available when a patron is looked for in the first place. Emigrants from villages with strong patrons tend to keep the home devotion if they keep any devotion at all. Those from villages without strong patrons are more likely to adopt the active patron of the town they move to, if there is one.

In order to present a composite view of how, on the individual level, communications with the divine and selection of personal patrons operate, I have selected several conversations with villagers as examples. They are presented in sets in order to establish by comparison the different criteria by which personal devotions can usefully be distinguished. It should become clear what kinds of people find God more accessible and how the different procedures for access to the divine operate. The few people presented in the next pages, then, are no random sample, though among them they probably represent the major varieties of religious stances adopted in the valley. In a sense the first two chapters of this study have been designed to provide an understanding of these conversations. The people I cite are without exception people I had been friends with for several months. The answers to the questions are given verbatim, but some questions and their answers have been entirely omitted from the transcripts. Note that in the excerpts from conversations transcribed here, the following conventions will be made: My questions, when transcribed, will be double indented and separated by a space from the response. When my questions are probes, and their nature can be deduced from the context, they will not be stated, but simply indicated by a slash (/).

VI. Gerónima, San Sebastian, Age 44

Gerónima is a widow with three children. She supports her family by operating a small store. At the time of the conversation (June 1969) she had just finished wearing the white habit of the Sacred Heart of Jesus and was waiting for the day of Our Lady of Mt. Carmel to don the appropriate brown garb. In both cases wearing the habit was the result of a promise.

In a different talk Gerónima expressed the conviction that the actions of Providence are to be seen all around, that the divine leaves signs for the knowing. In this she may be an extreme case, but it seems to run in her family. The signs she has observed are for her, demonstrations of the existence of God. She cited two cases. The first was the day of her wedding. On that day her wedding dress was lost, the only packet that fell from the burro coming up from Cosío, and she had laryngitis, the only time in her life. Afterwards she knew that these had been signs from God because her husband died only four years later and she was destined to be a widow for the rest of her life. (Widowed men remarry; women do not.) The week after her husband died she had a recurring dream. All she saw was the Sacred Heart of Jesus—"Just like in the church"—who held out his hand, took hers, and said not to worry, that He would be always with her. This happened 12 years before my conversations with her. She is gay, friendly, and articulate.

What devotions do they have here, do you have?

For me the most important is the Sacred Heart of Jesus, then the Virgin of Perpetual Help, and then maybe Saint Joseph, advocate of the dead. / Nobody here has any devotion to Saint Sebastian.

What about the shrines around?

Llanos, Salud, and La Luz, nothing more. I go to Llanos to make a sacrifice, for a promise, but not because I have more devotion to Llanos. The Virgin is one. I went to La Luz when I thought I was losing my eyesight after I had had my children. When I got there I thought a miracle had occurred, that my eyes were cured. I asked, "O Virgin, why have you granted me such a favor?" But then the problem came back and I realized I had just had a case of nerves. I have been to Salud every year except for a couple.

How old were you when you first went to La Salud?

I first went when I was younger than my daughter [age 12], with my mother. My mother always went. I have never gone there because of a promise. One time the day fell on a Sunday, so I went to mass here, and coming out of mass I felt a sinking feeling [points to her chest] so strong that I had to go. So I took off straightaway and arrived there and said three Hail Marys and came back. Another time there were people in the shop and so I didn't go. And I felt terrible, and I said to myself, "Are you not going just in order to make money?" I got all upset and tense, and I had to go. When I go I take my children, just as my mother would take me.

Do others go regularly from this village?

A few. Some years up to 15 or 20. But previously a lot more went. Another fiesta a lot of people would go to was the fiesta of the Sacred Heart of Jesus.

And when did you first go to El Llano?

Now they say that the Virgin appeared there, but I didn't go because of that. When I first went I had a promise to go. I went barefoot from Cosío. It was the year before the apparitions here [1961]. / Why there? Because it is a sacrifice. One time I went nine days in a row on a promise. I still have three promises to go there that I owe to the shrine. [Her daughter, at this point, said, "Let's go, Mama!"]

It must have been an important promise.

No—average. I didn't start going there until after my mother died.

And on what occasions did you go to La Luz?

The first time was with the picayos, hence not for devotion. Rather for entertainment. The second and third times were for promises. By then I was already a widow. The last time I said good-bye to the Virgin because I was so tired I thought I'd never go back. But I would still like to take the children there. My mother made a promise to go there barefoot if all of her children came back safe from the war. She was sick when they did, but she went anyway, stopping along the way to rest. What I enjoy doing is going, confessing, and receiving communion. This at Salud, Luz, and the Sacred Heart of Jesus.

What about El Brezo?

I would like to go there, as an offering to the Virgin. To see it, but on foot, like in the old days. I would not go as a sacrifice. There is more sacrifice involved when you close yourself in the church for a few Sundays in a row. . . .

When did you begin to take a serious interest in these Saints?

As soon as I could think. From when I was a little child.

Didn't you get bored by the prayers?

It's like my daughter. She gets tired when I make her pray with me, but when I go away, she prays by herself. So it was with me and my mother. / Devotion to the Sacred Heart of Jesus dates from the beginning. That is the most important, that is the Holy Sacrament. / Our Lady of Perpetual Help, that probably dates from when there was a mission here [1940, when Gerónima would have been 15 years old]. At that time I was still an infant. [There is a picture of Our Lady of Perpetual Help over the stove.]

Others?

I pray to the Guardian Angel every night before going to bed. When I leave the house: "Jesus Joseph Mary give us holy and good days." When I get up: Mary my mother, keep me from falling into mortal sin this day," then three Hail Mary's. When I start out on a trip, "Illuminate Lord all my steps. . . ." I made up all of these prayers myself. I like them much better than Our Fathers which we say without thinking about them, mechanically.

At what age did you begin making your own prayers like that?

Always. Take one time when I was in a quandary. I went to the Heart of Jesus in the house [it is in the dining room] and went on my knees, and I talked to Him, not praying, saying, "Look, God, this is weighing on me. Thy will be done, but let it be done one way or another so I can see my way out of this." and then I felt an enormous relief, and I felt white as the snow on the mountain-tops. The best prayers are those which come from the mind. I have another to the Virgin for her child, and whenever I am in an argument I say, "Aie, Lord, restrain my tongue."

Do you speak to Mary herself, or to her as an advocate to Christ her son?

Both. Sometimes I speak to her, saying, "Mary, look down upon us." And sometimes to her asking that she intercede with Christ for us.

And Christ?

To Christ himself.

What is the difference between Christ and God?

I never think about that. I know that we are told that Christ is the Son of God, and in another place that Christ is God. It is something that we cannot fathom. In the same way Saint Augustine was trying to figure out the mystery of the Most Holy Trinity, and he came upon a boy taking water from the sea in a pail. And Saint Augustine enquired "What are you doing that for?" And the boy said, "I am trying to empty the ocean." And Saint Augustine said, "But that is impossible." And the boy said, "What you are trying to figure out is more impossible." What he was trying to figure out was the mystery of the Most Holy Trinity.

What about the devotions of men?

A few men have gone on promises to La Luz. My father, for instance. Some youths, doubtless [cites some names]. My brother was an altar boy until he was 18. In the old days they would go to the front while the rosary was being said. He could recite all the vigils. And my son started serving when he was five. He knew all the responses in Latin.

What about Saint Anthony?

My father had a long prayer to Saint Anthony that he used to say in the Rosary every night. I can't remember it now. Father always started the Rosary. It would include:

1. St. Sebastian, patron of the village
2. The patriarch St. Joseph
3. The Blessed Souls in Purgatory
4. The dead
5. Our dead brother
6. The Pope
7. Any necessities of the moment
8. A Credo, a Salve, and an Angelus
9. The scapular of Our Lady of Mt. Carmel

It would take place in the kitchen. We would fall asleep. In November my mother would begin to recite the cycle of the birth of Jesus. That was a month long, an hour every evening.

Didn't that bore you?

Praying is a job, an obligation, like washing the dishes. I cannot say that I have always enjoyed it, or that when I do it I am not distracted. But it must be done, even if I am sick. It is very hard to say an entire rosary without being distracted; I am not ashamed to admit it. Especially in the long hours in the church during Holy Week, when I almost fall asleep. I even tell this to the priests—I keep wondering when it will be over. They say that those who say three Hail Marys every day with devotion will be saved. I don't always enjoy it. Sometimes I leave the church happy, other times sad.

What are the differences in attitude towards God of men and women?

Women are more timid, more foolish. They worry about whether they are praying enough. The men seem more open, untrammeled, and relaxed; they pray less.

Take your mother and father, for instance.

At night there was no difference, in the house. In the daytime there was. Father was off working, while mother could go to church. / No, father did not go to La Salud. But he was the majordomo when young. They had one family appointed that was the majordomo family. Mother washed the altar cloths. I swept out the church every day. Amalia would help out a little: she was responsible for the flowers. I was responsible for the side altars, and my sister for the front altars until we were married. I sang in the choir until recently. Our family was the majordomo family for about 15 years.

As one gets older, doesn't the nature of one's devotion change?

Of course. When I was young I thought much about God and I was enthusiastic about the Church, but I also thought about fiestas, boys, girls, clothes, and other things. Now we have more responsibilities, more problems, and we have to come closer to God. . . . I see the change in devotions that accompanies the progression from youth to old age as normal and natural, like a plant growing. And it is inherited from father to son. I myself want to die in the religion of my mother, and I want my children to be like me, if not better. When my mother died her last words were a simple prayer to the Virgin. She died with a candle in her hand. And my brother died with a prayer to the Sacred Heart of Jesus. They say that those who die with a prayer on their lips will be saved. . . . I would like to die the way my mother did. Those who do not die in the grace of God die like animals.

Why are we here in this world?

We're here to work. Christ himself worked. We cannot die with our arms folded. We must work to eat. And we must praise God, who gives us life and death and whatever comes afterward. One thing I am sure of. Those who have a full day's work will never be bored like the rich people who have nothing to do.

Some people worry over points of doctrine, or whether they will gain salvation. I don't see you that way.

No. I don't worry. I am tranquil. I think I will be saved. Furthermore I am consoled by what I read once, that those in mortal sin are restless, upset. As I am tranquil, I don't think I can be in mortal sin. Whenever I am upset I go and confess right away and receive communion. And I don't worry over points of doctrine. It is as if I owed someone a million pesetas. That is impossible to repay. I just do the best I can. The only thing I worry about is whether I pay as much as I ought to.

VII. Angela, San Sebastian, Age 43

Angela is a married woman with three children. Her husband is a herdsman of average means, and they live in the village of San Sebastian. She is open, friendly, and unaffected.

What shrines have you been to, and under what circumstances (this in many questions)?

The first time I went to La Luz it was as a teenager.

For diversion?

No. Just to go. Since then I have been only once on a promise, and that was for a brother who had been sick. My husband keeps telling me to stop making promises and praying for other people and save them for myself. I also went to Llanos for the trip (para ir), not on a promise. Oh, and once to Covadonga on an excursion.

You never go to these places on promises?

I suppose I have never had the need to. I have had few crises. / Yes, I suppose I would go if my children were sick, but so far I've simply had no occasion.

Where would you go?

To La Luz, not Llanos. Llanos is no sacrifice. It's not that I have more devotion to Luz, the devotion is the same; it's that it's harder to get there.

What about El Brezo?

I would like to go there once before I died.

To whom do you pray (many probes)?

San José when I get out of bed, the Sacred Heart of Jesus, and the Virgin. / No particular advocation of the Virgin. There is only one Virgin. / Above all to the Souls in Purgatory—novenas when there is any small crisis. And Saint Anthony—above all for problems with animals. Also Father Damián, a few novenas, but less. / No novenas to the Virgin. We have not had much suffering in our family yet. The thing I pray for most of all is that they follow the good road.

When did you start doing these various prayers?

When I was a child I prayed at night and in the morning, no more. The same goes for my children. I did not begin saying novenas until I was married. It seems that when we are married we have more problems. I have never worn a habit, as so many women here do. I have never had the need.

What prayers did your parents say?

They would always say the rosary, in the winter, at least. It would include Our Fathers for the Souls in Purgatory, personal obligations, Saint Joseph, the Pope, and others I can't remember. They always put me to praying first, I don't know why. Both father and mother took charge of doing the rosary. At dinner, Father said the prayers. Now at dinner I do—an Our Father to the Souls in Purgatory—nothing more. My mother always prayed to the Souls in Purgatory, put a five-peseta piece in their box, etc.

Do you ever talk to the Virgin, or Christ as if they were people, not in fixed prayers?

Yes.

What is your relation to the Virgin: sister, friend, mother . . .?

I speak to the Virgin as if she were my mother, and to Christ as if he were a friend
. . . as I am speaking to you now. / This is when I talk to them in church during mass,
not in a formal prayer.

What is the difference between the way you talk to Christ and the way you talk to
a friend?

When I speak to God, asking for forgiveness, revealing my faults, it is something else,
more concerned with the other world. I can be utterly frank, much more so than I can
be with you or a priest. I can trust him. It's not that I have bad deeds to reveal, it's the
bad thoughts. The mind never stops producing them. Every time I talk to Christ like
that, I come out of the church more tranquil.

What pictures of saints, etc., do you have in the house?

I am also likely to pause and say a spontaneous prayer to a Saint in the house while
I am doing housework—pausing in front of their picture. [The ones in the house are
Saint Joseph, Saint Anthony of Padua, Sacred Heart of Jesus, and the Virgin.]

What have you done to teach your children about these things?

I pray with them until they are about nine or ten, when I send them to say their
prayers by themselves.

At what age do boys start to pray less than girls?

As soon as I let them pray alone. I don't even ask my son [age 14] now whether he
has said his prayers.

If one of your children were sick, would it be yourself or your husband who
would make a promise for his recovery?

It would be me.

But if the child is the child of both of you . . .?

Yes, it should be both of us. We should be equally responsible. But we're not.

Why do women pray more?

Men pray a lot less, that is for sure. We pray more, and we are worse!

How do you mean, worse?

We are more boorish; we have wicked tongues.

Maybe that's because the men, being away with the cows all day, don't have
anyone to be nasty to whereas the women, stuck in the town, have too many people
to be nice to.

That makes sense, the men don't have the occasion . . . but it's not true, because
even when I go off down the road I often pray to myself. Maybe men do also . . . how
do I know what they do? I can't tell you anything about men. Except that even men
come in when they are in trouble. They are just as likely to put a five-peseta piece in
the Souls in Purgatory box.

VIII. Cecilia, San Sebastian, Age 50

Cecilia is a widow with several children. She is frank, somewhat bluff, with a good sense of humor. Her 16 year old daughter was present while we talked. She has moved away from the village since the time of the interview.

I began by going through a list of shrines. She had made one promise to El Brezo, another to La Luz, and had another one, as yet unfulfilled, that she had made to El Brezo. She had been to the ceremony of Our Lady of Mt. Carmel in Cosío, but only for the mass, not for a promise, because it is so close to San Sebastian. Her daughter had been to no shrines. Her father had been to El Brezo. Her mother had died young.

What Saints did your father pray to?

We don't know that. As the men are unlike the women—they don't talk about these things—I only know that he went to El Brezo a few times.

And you?

To the Souls of Purgatory and Saint Anthony. To the Souls in Purgatory, for sicknesses, and Saint Anthony, for animals.

How do you go about getting help from the Souls in Purgatory?

You ask for something, saying that if you get it, you will give so much to the Souls in Purgatory. You give it only after receiving the favor, in the box they pass around in the church during mass.

How long have you asked for things from the Souls in Purgatory?

Since early childhood. Since I could reason.

What about people that want to give money for other saints?

There is a little box above the baptismal font labeled "por el culto" that they put money in. They can go to the church and do that any time. I haven't heard the priest say this, but some women did, that it is foolish to have masses said for saints, when they are already saints. The Souls in Purgatory need masses said for them to get them out of purgatory. Notice that we are granted favors in the measure that we believe. If other people get things from Saint Anthony, maybe it's because they believe in him. When I ask for things from the Souls in Purgatory, it appears to me that they are listening, and they usually grant what I ask, but this would be because I believe in them. Other people might believe in other saints.

What about your children?

None of the children until they grow older have any faith. They would not go to mass unless I told them to, and they do not say their prayers unless I tell them to. Now, I'm not one to say my prayers every night either, really. When I send them to pray by themselves there is nothing but silence.

I have talked to some women who are very much caught up with the church . . . .

I am not like that. I go to mass, I go to rosary because I seem to enjoy it, but I don't pray that much. I pray for things, but I don't talk to God.

Cecilia was wearing the habit of Our Lady of Perpetual Help.

These three women present three varieties of personal religious stances open to women. They were chosen from the same village and in more or less the same place in the life cycle for the purposes of comparison. All have teenage children, all are middle aged, all are from San Sebastian. Angela is slightly exceptional, as she is married and the other two are widowed. As she herself says, she has not had many crises yet in her life, so she has not yet availed herself of the option to wear a habit or spend much time visiting shrines. Gerónima and Cecilia both have had hardships and problems, and hence have had bigger favors to ask, bigger promises to make to the divine.

The three were chosen because of the different role that religion plays in each of their lives. Gerónima is what Max Weber might call a religious virtuosa; Angela is typical in her religious stance of women her age in the village; and Cecilia is less concerned than most women with religion. Note that these differences do not appear much in the public devotions of the three women. All attend the rosary daily and mass every Sunday. But they clearly have different proportions of their life space taken up with God. For Gerónima the attention to the divine is very personal, very much thought out, at times even philosophical. Religion could truly be said to permeate her life, and in this regard she is like many of the more elderly women in the village. Perhaps because of her youth she is not as singlemindedly concerned with salvation as they are, but she matches their personal approach to the divine, their emphasis upon purification, and their attentiveness to the priest. Of the three women, Gerónima is the only one who could be said to have a personal patron, the Sacred Heart of Jesus. She may be exceptional in her ability to explain her relations with the divine, but there are other women similar to her in the village who have as complex a set of relations with as many different practices: an annual, regular trip to a shrine; devotion to a special saint in the village; the making of promises to other shrines; and the wearing of habits. She was clearly at ease talking formally about these matters, clearly on home territory. The other two women seemed to be wondering whether they were doing the right things and giving me the right answers. Gerónima had thought about all of these things before, and virtually had the answers ready. The other two were explaining what they did in religious matters to themselves as well as to me.

Angela matched the variety of Gerónima's devotions, but in her life religion played a less dominant role. This may simply be that as a person she is more relaxed, but it may also be due to her use of the souls in purgatory as a major source of practical assistance; it would seem difficult to have a personal relationship with the souls in purgatory, unlike the Virgin or Christ. My estimate is that Angela is typical of those women in San Sebastian who are devout but mainly preoccupied with their families and their work. Gerónima is no longer involved in the active farmwork of the village; this distinguishes her from Angela and Cecilia and is something else that links her with the older widows whose time is much taken up with God.

Cecilia was the most relaxed as a personality, and her attitude towards my questions and towards religion itself was almost bemused, much as that of many of the men in the village. Less at ease talking about her relations with God, she was more at ease with herself. For her, religion is almost a set of techniques, devoid of affective content. These techniques include all the major ones outlined above for obtaining a specific response from God in this world: promises of money, pilgrimages, and habit. But there is little or no personal communication with the divine. Perhaps this is because her mother died young and she was brought up by her father. Both Gerónima and Angela came from households in which religion had an important place; Cecilia's home seems to have been more relaxed.

Examining the social networks of these women, one notes that Gerónima and Angela are members of large families, with many relatives in the village. Cecilia came from a relatively isolated family, and she has no brothers, sisters, or parents left in town. Her nearest relatives are in the village of Rozadío.

Again it should be emphasized that these are, relatively speaking, very fine gradations of devoutness. San Sebastian is and always has been an exceptionally devout village, by far the most devout in the valley. Long before the apparitions it was noted for its adherence to ceremony and for its religious unanimity. The circumstances of life that provide such fine differences in San Sebastian might provide much greater differences in other villages.

Gerónima's religious life history gives a good portrait of a *core* parish family. The pattern is found in every village that I visited: At least one family, but usually a small set of families look after the religion with an almost proprietary air. They may be related to priests, like Gerónima, whose sister married a priest's brother. Or in the past the priest may have boarded with them. Although all villagers are literate, the core families are more literate than the others. In villages that have noblemen or manor houses the core families may also fill the jobs as attendants or servants. Their members might also be the ones (this is the case in two villages) who are the village poets, writing the verses of the picayos and keeping the folk paraphernalia for the fiestas. Also it seems that the core families are more likely to produce school teachers than others. The church is the center of culture in these villages, and the core families are closer to the church than most. In such families the father himself is likely to be devout, sitting under the balcony during mass. He would likely be among the more prosperous of the villagers and would carry himself with dignity.

At the other end of the continuum, the peripheral families are more difficult to categorize, for they may be peripheral to religion in a number of ways. One may be their relative poverty and lack of social standing. After a certain point a family simply lets appearances go. In the old regime these people were labeled as paupers and exempted from certain religious duties such as death masses. At present the abandonment of pretence applies to people in some of the other villages, but none in San Sebastian. Another characteristic of peripheral families might be enlightenment. Every village during the last years of the Republic had its *comité*, or Republican political junta. But rarely if ever in these villages was the comité antireligious. Nowadays in every village there might be one or two avowed anticlericals who stay away from the church, but their wives and children go. At present peripheralness is not a product of enlightenment, for generally education and exposure to the outside world bring more devoutness, not less. A third peripheralness is manifested in the quality of devotion. These people attend church and fulfill their public religious duties, but they regard themselves as outside the pale. This attitude especially should be seen not in relation to the Church *per se*, but in relation to the Church as epitomized by the core families and the virtuosi. If the core families receive the approval and the attentions of the priest and thereby set a village standard, the result is an alienation from the Church of the less active, less knowledgeable, less virtuosic practitioners. They fulfill obligations, but without emotional commitment; their heart is not in it.

Almost by definition, the nature of personal devotions of those near the core and those near the periphery will differ. Those of core families will be up to date, and like Gerónima's they might include those of the Sacred Heart of Jesus and Our Lady of Perpetual Help, devotions propagated in the lifetime of the villagers by priests or missionaries. The core families are precisely those families closest to and most influenced by priests. In contrast the peripheral families' devotions are more likely to be those that were core devotions in the previous century, or even before. Like Cecilia their chief spiritual succor might be the souls in purgatory, the chapel devotions, or some other devotion not spread by the priests. Naturally there will be much overlap, especially when the village in question has a strong, active patronal image, like Lindes, Llano, or Vado de la Reina. Then attention of core and peripheral family alike centers on the patronal image to the relative exclusion of other images. But on the whole the core families will be more up-to-date in their devotions and will accept wholeheartedly the brand of religion presented by the priest. In contrast, the peripheral families, like Cecilia, will see the religion from a practical point of view, one that is probably common in its essentials to the village for centuries past, if not millenia.

Perhaps the essence of this core-periphery distinction can be stated in terms of the villager's perception of her place in the society. The daily rosary, attendance at mass, and support of the church are village duties. They are part of the corporate responsibilities of the villagers toward their God. To the extent that a person considers herself an important member of the village, then she will feel a responsibility as a villager to fulfill the amenities in regard to God. So to a certain extent status is involved—perceived status in the village. This of course is related to the closeness to the priest and schoolteacher, because these are the twin poles of status in villages with no great differences in wealth and no noblemen.

Two men from Tudanca provide another set of contrasts in styles of personal devoutness.

IX. José, Tudanca. Age 42

José lives in a town in central Santander, but often visits Tudanca, where he was born. His father was a sawyer. His mother and unmarried sister, deeply involved in church affairs, both live in Tudanca. His family has always been closely connected with the manor house. His uncle worked there, as did his sister. In the process the family became associated with the family of the manor—the kind of subordinate kinship discussed in the first chapter. The nobleman used to bring notables to the village, even Cabinet Ministers, and drop by José's mother's house. It was the most old-fashioned house in the village, and there they would "sit on a stone and drink wine and be happy."

José worked for many years as a miner, a tunnel laborer, but for the last 13 years he has been laid off because of silicosis, rock powder in the lungs. He is dangerously ill and cannot do any strenuous exercise. He can, however, get up and walk around. He has plenty of time to talk and is thoughtful, articulate, and knowledgeable. Since the conversation I had with him was very long, and I did not take notes, it is not recorded in the form of a dialogue.

Both his father and mother had as their prime devotion Our Lady of Vado de la Reina. In addition, his father venerated Our Lady of Covadonga and Our Lady of Pilar, both of which he visited while working as a journeyman sawyer. His mother was devoted to El Brezo. His sister's prime devotion is to Vado de la Reina, with secondary attention to Saint Anthony. She has worn the habit of Our Lady of Mt. Carmel and has visited a number of other shrines on promises.

His devotion to Our Lady of the Snows dates from infancy. "I don't know whether somebody put it down my chimney, or it was because I just copied what others were doing." But it was not until the age of 14 or 15 that he began to think about things and seriously took the Snows as his favorite. In childhood he had followed the devotion because "it was the only thing in front of me. Later, when I began to think, I saw that there were others I might choose, but chose this one."

"People always have one devotion more concentrated than others they might have. In Santotís most have devotion to Our Lady of the Snows, some to Our Lady of the Vega, but always one is more important than the other. Children usually (but not always) take on the devotion of their parents."

When he was drafted into the army, his sister bought him a subscription to the magazine of Saint Anthony, *El Santo*; when it ran out he renewed the subscription, and he has ever since. So the devotion to Saint Anthony provides a kind of secondary devotion. But his prime devotion to the Snows remained. "When I was drafted my family offered a promise for me to carry the statue all the way from the shrine down to the church should I come home unhurt. When I came back I thought it was a good thing, so I did it myself. I have done it on two occasions, once up, and once down. I don't say there aren't others more devout than me in most things, but as to devotion to Our Lady of the Snows, I cede to no one."

Now that he lives away from Tudanca he has just as much devotion to the Snows as anyone. "Maybe more. It sometimes seems to me that I have more, because I am away and cannot get back often. It is as if she were more capable of helping me. Not to cure me exactly, but maybe to help me suffer less, to console me, to make me stronger. It seems to me that she can do this better than other Virgins. That she is simply more powerful.

"The shrine that people go to where we live is one to Our Lady of Mt. Carmel. On the fiesta day there, there might be up to a thousand candles burning—masses from seven in the morning all through the afternoon. Previously many used to go barefoot, now fewer. We go to that shrine, because we love the Virgin, but it is not at all the same thing as the Snows. My wife feels that way, too.

"We prefer to go to the fiesta of the Snows at Anievas, nearby. That seems to me to be almost the same thing as the Snows here, although it is in a different town on a different day. It is a small fiesta. It would be better to be here, but it seems to me to be the same devotion."

His daughter, aged 10, shares her parents' devotion to the Snows. Last year she wanted to go all the way to the shrine on the day the statue was taken up from the village. He does not feel that in this respect his family is exceptional; he thinks that those who had devotion to the Snows before they left the village maintain it while they are away.

José very much wanted to attend the shrine ceremony on August 5, and he said he would borrow someone's horse if he could be in town. Because of his delicate health, his family had the priest say masses to the Snows for him. When on the previous Sunday he had heard the priest announce that the mass on Thursday would be dedicated to Our Lady of the Snows for the intention of José Gonzalez he was so moved that he wanted to weep.

I asked him what relation he thought God had to problems that we encounter.

"Some say that God wants us to suffer, so he puts trouble in our way, like your war in Vietnam. They say we need to suffer. This does not seem right to me, not reasonable. It seems rather to me that God is simply not involved in our everyday activities, either because he chooses not to be involved, or because he does not have the power to act. I lean toward the latter explanation. God's main activity comes at the last judgment. We play our roles while we are alive, but when we die we must have our accounting with him. God is . . . above."

The Virgin, on the other hand, "actively helps us, whether this be by intervention with God, or whatever. She is closer to us, does things for us in this world."

The Devil "is always at work, trying to turn us off the right path, trying to get us to do wrong."

He said he thought that all the people of Tudanca, and all of Spain, for that matter, would go to heaven—maybe a few to purgatory—"if what they tell us is true." He said that the less devout, in these villages, generally had parents who did not teach them about religion, that it takes a long time before one can piece together into a meaningful whole the things one learns by rote in catechism.

José is ambivalent about priests. Indeed, when I first brought up the subject of religion, his reaction was a diatribe against priests. "Not that I'm not good friends with them. But they butt in where it's none of their business—like what I do with my wife—in confession." He confesses, but doesn't like it. He told the story of being chummy with a Dominican at a bar, then going to confessional and being blasted by the same priest for what they had talked about in jest. Like many men he resents them.

When I asked him about differences in devoutness between men and women, José maintained that there was no difference. "True, women do go to rosary. But it seems as though the rosary is by tradition a devotion for the women. Similarly to go on the final part of the trip with the statue to the shrine. It is not that they are more devout. Men used to go more, but the custom of men going has lapsed. When masses are ordered it is the woman who handles it, but that is like when you send a kid to the store to buy something for you, it is really you who buys it."

X. Miguel, Tudanca, Age 53

Miguel is a herdsman of average means.

I find myself interested in what people believe in, and that's what I've been studying here. What about you?

Nothing. God? There is something . . . (algo habra). The Virgin? Just as little. After death they can throw me in the bushes. There is nothing after death. Until something terrible befalls them, they don't do anything; until a fall, an illness, or little before death—yes, then; but up until then, nothing; and like me the rest of the men here.

I go to mass; I like to go when I can. But when I can't, I don't. You have to wash up and dress up, and then change clothes back again before you can go off to the cows. Mass is for those who can afford to go and calmly have a drink afterwards, then go home for lunch. I tell my wife that I don't want to bother anybody about it, and that nobody should bother me about it.

As for confession I confess every year, as promptly as anyone, but I tell the priest what I want to tell him. What right does he have to know my secrets? My wife knows this. She lets me go my way, I let her go hers. We respect each other's ways.

Were there moments in your life when you believed more than you do now?

Yes, up to, say, age 15, when I turned bad. Afterwards: nothing. And the priests? Don't talk to me about the priests.

Are there other men who believe more than you?

Some act as though they do. Juan for one. But he would do you in if he had the chance *(fastidiarte)*. There are those who go more to mass. But that does not mean they are more Christian. I compare my wife who goes to mass on Sundays and doesn't bother anybody, to those women who come out of mass, confessing and receiving communion every day, and then do their damndest to take what they can away from you, when they get a chance. Or to Sylvia in the next village, who is always tearing everybody down with her sharp tongue. That's why I don't go to mass more frequently; they are hypocrites. Going to mass doesn't mean anything. Sylvia, who is a friend of the priest and always ordering masses! It's just as likely that those who don't go as often are better Christians than those who do. Acts, not appearances, are what count.

Were your parents devout? Did they say the rosary in the family? Do you?

Father was very devout. Every evening we would say the rosary in the family; then getting in and out of bed we would always cross ourselves. When I got married the wife got us to say the rosary in the evenings. But then for one reason or another we would miss it, like being tired, and finally I said, "let's not be ridiculous" and we dropped it altogether. This in spite of the fact that the magazine *El Santo* and the priest are always talking about devotions within the family. This is the same for everyone in Tudanca. I would wager that not one family recites the rosary now, and in the entire vale very few.

In order to make José and Miguel more comparable I have chosen them on the basis of similar age and position in life and because they are associated with the same village. They represent two clear types in the villages; the one sincerely devout, the other frankly sceptical.

José's family is a prototypical core family, and he is a part of that milieu, even though he lives away from the village. Being laid up for so long with so much time to himself he has had time to formulate ideas and nurture his affection for Mary. For it is that relationship of deep reverence and affection for the Virgin of Las Nieves which forms the central aspect of his religious experience. His affection, as he says, is sharpened by his absences from the village. This is the case with many emigrants. For instance, while only two or three women in the village were given Nieves as their baptismal name, it is given very much more frequently as the baptismal name of children of emigrants to other villages.

In addition to coming from a core family José shares another characteristic with Gerónima. He is an isolate from the village work. This gives him time. Time does not necessarily lead to devoutness, but for those who are devout time helps to develop formulations of devotion.

Miguel is not originally from Tudanca, and this may help to account for his lack of interest in Our Lady of Las Nieves. But in fact his is a disinterest increasingly shared by other men in the village, as he intimates. What Miguel said in this conversation many other men in the village had been reticent to tell me (they considered me to be devout because I was interested in religion). From Miguel's case it would seem that this abandonment of commitment to personal (as opposed to public) religion has taken place in this generation. But because of the relative silence of men on these matters at all times in the village history, it is difficult to say what they thought or felt in the past.

How typical are José and Miguel? I would estimate that out of 80 men in the village, five or ten had views and religious experiences similar to those of José, and 20 or 30 were similar to Miguel. The rest would fall in between, with a fairly thorough disinterest in institutional religion, but with the maintenance of some personal devotion to Las Nieves. Almost all of the men, in spite of the lapse of personal devotions and institutional disinterest, maintain the traditional forms of propitiation and promises.

Both are atypical in having clear and well-formulated viewpoints on the matter. Discussions of this nature about religion rarely if ever occur. The mere fact that José and Miguel, while having such completely contrasting attitudes to the divine, could see all the other village men in their own image testifies to the lack of discussion of beliefs and devotions among the men. This is a very private matter for men, shielded even from their wives. Their positions are kept inscrutable for all others, including especially the priest. It may be that men's opinions on religion are kept inscrutable even for themselves. As a general rule, things that are not talked about do not need to be thought out. There is no need for a commitment either way; the mind has no particular call to make one. This is, perhaps, the reason for silence in the first place—in order to leave a margin of freedom in a time of changing values.

What aspects of belief and practice do José and Miguel share? How in these respects do they contrast with Gerónima, Angela, and Cecilia? What can these five cases tell about differences in devotion and practice between men and women? First of all, both José and Miguel leave the technical aspect of negotiation with the divine to their wives. In José's case this is merely a proxy situation, a product of the division of labor in the household discussed above. In Miguel's case it is because he himself is manifestly disinterested in the whole thing. In both cases the relegation of practical chores (masses, promises, etc.) to the women shield the men from public scrutiny of their positions. This too helps explain the radical difference in José's and Miguel's perceptions of the rest of the men in the village. The public devotions of the town women are fairly unanimously observed as a matter of custom, and their participation is always open to the interpretation that they are acting as family proxies with the active agreement of their husbands. Even the women do not always know just to what extent their religious activities on the behalf of the family carry their husbands' active approval. Witness the statements of incomprehension that Angela and Cecilia made about men—"How do I know what they do? I can't tell you anything about men"; "Men are unlike women—they don't talk about these things."

José, in spite of his devoutness, shares Miguel's hostility to priests. The women, on the other hand, mention priests, if at all, as authorities and as models for behavior. The men's hostility, fairly universal even if generally unspoken throughout the valley, may also help to explain the use of women as intermediaries between families and the Church—they are acting as buffers between the men and the priest. In this light, José's analogy of sending a child to the store on an errand is seen to be imprecise. A man or woman often sends a child to the store on errands, but on other occasions they go themselves. Men almost never transact religious business directly with the priest. What lies behind the male hostility?

To begin with there is the reason that José gave—that the priest interferes with matters in the confessional that are not his concern. Indeed, much of the hostility toward the priest involves confession. As a shepherd described it, it is an annual problem. "The woman tells the man that the time of year for confession is approaching, and fixes up his clothes, and sends him off." An older woman described it the same way, "We always have to push the men to get them to go to confession. They say that they have this and that to do. They don't want to go." Hence women are not merely neutral buffers. In their role of monitors of the family's relations with God they must, at times, be out and out allies of the priest. The resentment toward the woman-priest axis is clear from Miguel's angry reference to Sylvia from down the valley—"Sylvia who is a friend of the priest and always ordering masses!"

The priest and the church he represents are in competition with the men for the time of the women. In a sexist society, women are resources. Most practically, they are active hands and arms, usually necessary for the operation of the family enterprise. As the administrators of the enterprise, men are sometimes upset to see their labor straying off to the church.

This interference in the working unit parallels an interference in the sexual unit. José resented the priest telling him what he should do or not do in bed. And although the doctrine of the Church has long supported the male's sexual dominance and virtually enforced copulation, the production of large families in rural European society is no longer seen by the people themselves as desirable. Because of the universal conviction that the society and the way of life are dying, children are no longer seen as hands to help, but rather as mouths to feed and bodies to clothe. While in the past the priest served the farming family, in a sense, by enforcing its fertility, he is now standing in the way of family prosperity.

Not only in the doctrine that the priest espouses but also in the mere fact that he intrudes at all into the sphere of the family unit, he generates hostility from the men. For in these villages, as we have said, there is a great emphasis upon equality and the independence of the vecino unit. All units are equal, all are separate. Confession regarding matters within the unit challenges the autonomy of the unit. This same point comes clear in the resentment of the priest's traditional social alliances. Predictably, given his education (though not at all logically, given his creed), the priest associates with the thin layer of rural elite in the valley. These include the noblemen, when there are any around, the proprietors of the more prosperous commerces, the officers of the Guardia Civil, and the professionals. The social intercourse of this stratum is largely limited to card playing in the cafes, where regular games take place virtually every day. The priest also has political roles in the vale councils and other government bodies. His *de facto* alliance with the government and the local power structure is one more strike against him. For the government itself periodically interferes with the democratic process of village government and the autonomy of the vecino unit.

There are still other reasons for the hostility. The priest, by offering an alternate to the authority of the master of the house, threatens the unbalanced equilibrium of the household. Personally, he is generally not a sexual threat. (In fact, the value of the religion in keeping their women in line was mentioned time and again by the men.) But in the attention that he gains from some of the women the priest is certainly an ideological threat, an authority whose opinions might conflict with those of the husband. He is an alternate model.

One manifestation of the masculine ill-will toward priests comes out in endless stories about the priests' ambiguous sexual position, tales of priests living in sin with housekeepers, and jokes referring to priests' emasculation. In a sexist society, the priest is at a remarkable cultural disadvantage. The new, younger priests, who have a more liberated attitude toward sexual matters, are closely watched by the men. In the next valley a young priest was severely beaten when he was found with a girl.

This kind of tension must be seen in perspective. While in many ways the men resent the priests, there is little or no questioning of the necessity for having them. Except in very rare cases, the men treat the priests with the most punctilious respect and have protected them in time of danger. None of the priests in the valley were killed by the villagers during the Civil War, and many were hidden for varying periods of time in houses. This is a deeply religious culture, and the priests are indispensable. Hence the criticisms are directed toward a particular priest or priests in general, but not toward the role of the priest. And exceptions are made, as in the case of a saintly young priest who lived in Polaciones, Miguel Bravo. Villagers of the valley say that Miguel Bravo would give anything away that he owned, that he would be ready to help out his parishioners at a moment's notice, that he served his parishioners.

The differing attitudes toward priests is symptomatic of a deeper divergence in religious attitude between men and women. The great, fundamental difference between the religious stance of men and women lies in their approaches to the notion of sin. Gerónima expressed the essence of this difference when she said, "Women are more timid, foolish. They worry about whether they are praying enough. The men seem more open, untrammeled, relaxed; they pray less." Although the religion enriches some aspects of their lives, psychically it oppresses the women, while leaving the men relatively untouched. Even with all of José's affection for Las Nieves, his life space is only very partially taken up by religion. In spite of his being an extreme case among males in the village, his attention to religion comes nowhere near matching that of Gerónima or Angela. This difference involves a fairly complete lack of anxiety on his part. How does it come about?

It seems likely to me that behind the multiple popular devotions and religious activities of the women lies a sense of impurity that has been laid on them by the Church and its ministers. Ever since Eve, women have been seen as temptresses, slaves to passion, the cause of man's downfall. With an overwhelming emphasis on female models who were pure and virgin, with villagewide membership in such puritanical organizations as the Daughters of Maria, adult, sexually active women cannot help but feel that there is a sense in which they are unworthy, that they have something to expiate.

This attitude has been somewhat cultivated by the Church in priests. A handbook for priests printed in Madrid in 1924 is worthy of quotation at some length:

> What is a woman? Saint Jerome replies that she is the door of the devil, the way of iniquity, the sting of the scorpion. In another place he says that woman is flame, man tinder, and the devil bellows.
> Saint Maximus calls the woman the shipwreck of man, the captivity of life, the lioness that embraces, the malicious animal. Saint Anastasius the Sinaite calls her the clothed serpent, the consolation of the devil, the office of the devils, the ardent oven, the spear in the heart, the storm in the household, the guide of darkness, the teacher of sins, the unbridled mouth, the calumny of the saints. Saint Bonaventura says that the woman dressed up and lovely with her finery is a well-sharpened sword of the devil. Cornelius Alapidus says that her look is of basilisk, and her voice that of a siren; that she bewitches with her voice, befuddles the judgment with her look, and with both things corrupts and kills.
> God grant that our experiences do not confirm these observations.[54]

This passage is in a section of the pious booklet designed to warn priests of the dangers of contacts with women. In the process much of the misogyny latent in

Church doctrine comes out into the open, for the concept of woman as Eve predominates. Note the symbolism of man as tinder, woman as flame. The burden of guilt for the expulsion from the Garden is placed on Eve; man was a passive victim, burned by the feminine flame, which is why the burden of impurity and sin falls on women, not men. It is put on the women by a male-dominated culture and a male-dominated Church.

The entire sacramental hierarchy is made up of men, including the Pope. The only women granted any measure of status are nuns or other virgins who become saints. While Church doctrine speaks of the equality of man and woman before God and sacramentalizes the married state, by the very structure of the Church and the implicit sense of the doctrine, married women cannot help but feel inadequate—first because they are women and second because they are married. The implication that sex is somehow less holy than celibacy is inescapable. Why else are the priests and the religious celibate? Why else does one lose membership in the Daughters of Mary upon marriage? Is this not one of the reasons that little children are considered to possess a sacred innocence?

The churching of women after childbirth is another instance of the way that women are made to feel impure. Technically this is an act of thanksgiving. Theologically speaking, "No idea of purification whatsoever is contained in the rite, for in child-bearing is incurred no sort of taint . . . . The common idea that the mother should not go to church for any purpose before being churched is a pernicious superstition."[55] A pernicious superstition indeed, pernicious and persistent because the ritual itself, and the name it is given by the Church are so clear in their meaning. The second of February is entitled the "Purification of the Virgin." This was the day that Mary was churched. The symbolism of the churching ceremony hardly betokens thanksgiving. "The woman kneels at the end of the church holding a lighted candle; the priest sprinkles her with holy water and recites Ps. xxiii, and then leads her to the altar rails . . . ." The ritual has the woman being first purified with holy water, then escorted back into the most sacred part of the church by the priest. The Church may change what it says it means by the ritual, but the ritual itself speaks loud and clear its own meaning. The fact that popular belief clings to the notion that women are impure until churched testifies to the clarity with which the symbolism is perceived.

Another element at the root of the feminine sense of unworthiness relates to the distinction made in the first section between the village as the female sphere, with its quarrels and estrangements, and the masculine sphere of the countryside. Because the women are concentrated in the village, their chances for stepping on each other's toes are much greater than are those of their husbands out in the meadows or the winter barns. Certain standard sins like covetousness and bearing false witness are virtually inevitable in the course of a day in the village. As Angela said, "Men pray a lot less, that is for sure. We pray more, and we are worse!"

"How do you mean, worse?"

"We are more boorish [villana], we have wicked tongues [malas lenguas].

In the context of this sense of unworthiness the shrine or the personal patron has a dual role in the life of the average woman in the valley. It is the site she repairs to in time of crisis, either to seek aid or give thanks, and it is also a place for a regular spiritual refreshment whether annually or more often, a place where all sins are washed away, where the strife and rancor of daily village life is distant, where she renews her sense of purity, sharing (as is most often the case) in the purity of the Virgin. For the women the shrine is a special high-power redemption center. Gerónima, speaking of her colloquy with the Sacred Heart, said that afterward she felt pure as the snows on the mountain top. And using the same image as Gerónima, the priest who spoke of the shrine as a cleansing station: "Life is like a road, but on the road we get dirty. Coming to the shrine we come to a place where we can become clean, clean as the snow of Our Lady of the Snows."

In what I have seen of Catholicism in the Nansa valley, much of the mental energy that goes into involvement with the divine concerns cycles of purification. The cycle already described—the ritual cleansing of contact with the workaday village world, sex, etc. of adult women—involves the shrine as a purification station, the operant cleansing being done through the powers of forgiveness of an all-knowing, all-understanding mother. The sin—the pollution, in the case of many women—has been internalized. The entire process is emotionally charged. There is a real engagement within the religious system, one that is relatively unmediated by priests. For many visits to the shrine image the priest is not present. In these moments of great religious intensity the contact between the personal patron and the person is a direct one.

This cycle in the married woman is similar to other cycles of purification, on other scales, in a person's life. On a smaller scale in a smaller time period there is the sacrament of confession and communion and penance, which is at least officially a periodic purification. For children this might be the first Friday of every month; for some men it might be once a year; for other men it might be the four or five major feast days; and for some women it might be daily. Confession, communion, and penance are sacramental acts that are commonly coupled with visits to shrines, especially if the visits include promises. In these situations, while it is the chaplain who grants absolution, the people feel that it is the image, or the divine figure it represents, that really secures for them and dispenses the necessary grace.

The cycle for women of cleanliness, pollution, and purification is related to a larger cycle which includes the entire life span of virtually all persons, male or female. The latent concept of woman as Eve relates to a notion of the adult world of labor as the place that Adam and Eve were expelled to from the Garden. Just as the Garden of Eden was an idyllic site, free from care, so the Spanish villages treat children as exempt from many of the more onerous duties.

The reason, of course, at least in the earliest years, is the ineptitude of the children for major responsibilities, but this exemption is accompanied by an attribution of innocence. The first communion ceremonies every year at Corpus Cristi provide the ceremonial apotheosis of the concept of childhood purity, The children are dressed in white and treated as if they were angels. It is irrelevant that these are the same urchins who the week before stole into several homes to pilfer during the church service. At that point they were innocent; they had not yet heard the Word and rejected it, so they did not know what they were doing, and they were not severely punished. The priest expressed their prototypical role in his sermon, "These children who are to receive communion are the best persons in the congregation. They represent purity and innocence. They are an example of what we should be."

It matters not, ultimately, whether the children are really innocent. They are fulfilling a need for the adult generation. For the adults the children are inhabitants of the Garden; they form an idealized subset of the village into which the World, the Flesh, and the Devil cannot penetrate. The week before Corpus Cristi in 1969 was the last week in May. Then, following tradition, the children of the village present the Virgin with roses to mark the last day of May, considered to be the month of the Virgin. On this occasion also the priest recommended the innocence of children to the adults as exemplary (as if that were possible! for the children's innocence, to the extent that it existed at all, was one based not on lack of sin so much as ignorance of it. Adults have no more option than Adam and Eve to abandon moral knowledge). The children were embarrassed and frightened by the ordeal of the presentation of the flowers, in which each had to make a public declaration to the Virgin in front of the entire village in the church. It was a ceremony written and staged by adults who would not or could not say those simple statements themselves. In a religion that centers early and often on purity, young children are made living ideals of the way to be, as if to prove purity possible and remind adults that while they are no longer pure they hold the potential or germ of purity. It is coincident with these notions that the younger the children are (and therefore the more innocent), the better they are dressed, the better they are fed, the more they are washed and attended to. Over half of the images in the parish churches of the valley include a representation of the infant Jesus. The day-to-day life in these villages could be called child centered because it is purity centered.

But gradually the children, because they are growing up, must leave the Garden, take up chores in the fields, and join the more profane state of the rest of the village. Whether as a cause or as a result of this growth, they must lose their innocence. The process of losing innocence and gaining moral knowledge resembles the Fall from Grace in the Garden of Eden. (That is one reason why retarded adults are treated specially in these villages, for in not growing up they

have not lost their innocence, they are not considered to possess moral knowledge.) As children from age seven to thirteen work more and more in the family
unit, they are also under the guidance of the schoolteacher and priest—learning
catechism, confessing and receiving communion rather more frequently than the
rest of the village, being taught to understand sacramentally the meaning of the
changing condition of their bodies and their tasks.

Intrinsic to the transition period is experimentation with moral boundaries, the development of a personal sense of sin, and the resultant formation of
a moral identity. As we noted before, women seem more vulnerable to the
internalization of a sense of sin than men, probably because they are taught to
be affectionate and sensitive to emotions from early childhood, also because
they generally accept the attribution of their being somewhat unworthy and
intrinsically sinful which the Church and culture manage to communicate to
them.

The final step of the descent from the mountain, the irremediable exit
from the Garden of Eden, is marriage. In San Sebastian the mothers cry when
their daughters are wed. I asked my widowed landlady why. Her response was,
"Because they know in advance what kind of life the couple is in for. And it will
not be good. Crosses, Crosses, and more Crosses." Adam and Eve were expelled
from the Garden to live by the sweat of their brow. Eve had to suffer the pains
of childbirth, and even Jesus had to carry his Cross before the Resurrection.

In the active years of the life cycle a tension is apparent between the
concepts of human nature as dominated by self-interest and human nature as
devoted to the service and glory of God. It is precisely the necessity to pay
attention to one's own needs, sometimes almost exclusively, as demanded by the
system based on private property and the land tenure, that helps maintain (along
with sex and uneven tempers) the notion of the profanity of the active life. The
dichotomy between the sacred village and the profane world (discussed below) is
replicated within the village by the dichotomy between the sacredness of
childhood, priesthood, and (to a lesser extent) old age versus the profanity of
adulthood. Feminine impurity is an intense subset of adult profanity.

In the religion of the devotional handbooks the profanity involved in
worldly activity is implicit, given the bias toward contemplation and meditation,
toward the goal of unison of soul with God. Time spent on oneself or one's
family was time taken away from God. St. Francis, St. Teresa, and St. John of
the Cross encouraged a selflessness, as did the nearby alternative life style of the
monasteries. Now, under the new doctrines of the younger priests, the same
point is made in a new way. Activity *per se* is no longer valued less than
contemplation. Indeed, fruitful activity is generally considered to be more useful
than contemplation by the younger priests. But only certain types of activity are
considered to be of transcendent value, those that aid others. Since most activity
must inevitably be for oneself or one's family in these villages, the villagers are

left as distant from sainthood as ever. (Indeed, more distant. Because under the contemplative system, one could always redeem oneself in one's old age. But now the old folk are left out of the kingdom, for there is little useful activity for others that they can perform.)

The result of the classification of self-interested activity as profane is that either the active villagers internalize a sense of impurity and spend much mental energy and, indeed, anguish upon redemption (this in spite of the priests, who are bored by most of the confessions they must listen to), or they dismiss the religion as either temporarily or permanently irrelevant to their condition. The latter either come back to it when their condition changes, much as the Indiano reconverts when he returns to the village from his years abroad, or they do not come back until the deathbed. The former (either a continued internalization of impurity, or a temporary dismissal of religion until widowhood) is generally speaking of the situation of women. The postponement of redemption in everything but a sacramental sense until the deathbed characterizes many of the men. The men see the religion as degrading, and being taught and expected to dominate and lead, they find it difficult to reconcile with their dominance the submission, the confessions of inadequacy, and the invasion of privacy demanded by the doctrine of the Church. Also involved, generally, is a sensible recognition on their part of the impossibility of the ethical demands made upon them.

One obvious weekly reminder to the men of the relative profanity of the life of toil is the precept, previously rigidly enforced, prohibiting work on Sundays. While exceptions were always made for haying in August, also so that neighbors could work together for poorer or short-handed families, the church records as far back as they go show a constant struggle against the centrifugal tendencies of the herdsmen. In every village this remains a point of tension. It was one of the first that Miguel made, and he made it with some vehemence. At present the specific norms vary from village to village, but in all villages it is a friction point symptomatic of the clash between two modes of life.

Behind the words of Miguel describing why he does not go to mass ("Mass is for those who can afford to go and calmly have a drink afterwards, then go home for lunch") may be the notion that activity and religiosity are incompatible. Or, when they are incompatible, that work takes precedence over devotion. This does not mean that Miguel experiences a sense of sin or impurity for working on Sundays. His position, and that of many men in this valley, is more akin to that of the Sinhalese peasants as described by Gananath Obeyesekere:

> Villagers often freely confess: "we are sinners." This has nothing to do with any doctrine like that of original sin, but rests on argument such as the following: "He who has been born as a poor peasant must have sinned in previous births. But he who is a poor peasant is compelled to perform actions that are sinful—hunt, fish, destroy insects, reject the request for alms by a beggar, etc.—in order to make a living. Thus all

poor peasants are sinners." ... This use of the word "sin" is not obsessive in character; it is used in a matter-of-fact manner and illustrates the point that sin is built into the context of social life and is its inevitable counterpart.[5][6]

Durkheim distinguishes impurity from profanity. Impurity is a kind of negative sacredness, while profanity is, as it were, the condition of the secular world, apart from either impurity or sacredness. This seems to me to be a useful distinction to make here. Miguel and others of his mind are different from the women and (see below) the emigrants. The latter experience a disquieting sense of impurity. Miguel experiences only a sense of profanity, which disturbs him little. The impurity of the women and the emigrants must be exorcised by rituals of purification, the visit to the shrine, the receiving of communion, the wearing of habits. The profanity of Miguel is not his fault. It is ascribable to his condition as a working man. He is not moved to do anything about it. Miguel's state is not a violation of the sacred system so much as it is outside the system, beyond it. He is in a different world, and his conscience on this matter cannot be touched by the Church.

To make this distinction is to raise again the question why women feel impurity and not merely profanity. Surely they could find "outs" similar to those of Miguel—such as their biological condition as women or their necessary role as producers of children—that would free them from a sense of responsibility for their condition or their acts. Conversely one could ask why men do not experience a sense of impurity from their sexual activities.

The simplest answer is that I may be overestimating or exaggerating the degree to which the religious activities of the women are indeed motivated by a sense of impurity, a need for cleansing. My judgment on the matter may be faulty. The images of purity used by the women to describe their feelings after communion, prayer, or visits to shrines; the doctrinal bias against women implicit in Church structure and myth (woman as Eve); the emphasis upon the sexual purity of Mary, the paragon of womanhood; the almost universal taboo on menstruation; and the ceremony of purification or churching of women after childbirth, all would tend to generate in women a sense of impurity. Yet this is a judgment on the state of mind not merely for which I have little direct evidence, but also which, if it exists, would seem to be only half-perceived by the women themselves. It is undoubtedly true that people can act on the basis of a feeling that they do not understand (as in the case, "Why, I believe I must have been jealous when I did that!"). But it is difficult and presumptuous for others, especially of the opposite sex, to uncover their motives for them, and I wish to express here my uneasiness and hesitancy at so doing.

Perhaps a simple answer to why women are made to feel more impurity than men has to do with the usefulness to the male-dominated society in having women subject to a self-enforced sense of unworthiness (to keep them as a docile labor force and maintain the male honor by assuring female fidelity) and the consequent barrage of models and rules that women are taught from childhood, leaving them, in fact, no conceivable "out." Why the men are allowed to avoid a sense of impurity would be the converse of the situation of the women. The men run the society; the men make the rules; the rules are easier on the men. Not coincidentally, the Garden of Eden story, which seems to be the ideological charter for this kind of society, has Adam less guilty than Eve. (Also, one should note, it explicitly awards Adam, after the expulsion, dominance over Eve, and symbolizes this award by having Adam *name* Eve.)[57]

The women, unlike the men, usually come back to the church with renewed determination as soon as they can spare time from their families and work. During the period of her maximal activity, when she has young children in the house and the husband is away on spells woodchopping, the woman's assiduity at some of the public devotions like the rosary or the morning mass necessarily declines. But when her children grow older, and especially when her husband dies, she has more than enough time to make up for it. It is then that she may become one of the elderly ladies in black that are seen haunting the church in every European village, for the death of the husband is a change in the life stream equivalent to that of passing out from the shelter of parental protection, and as in that earlier situation, a turn to the church is a solution to a sudden sense of aloneness. Then too there may be the feeling that some kind of reparation is necessary for the relative lack of religious activity during the busy years. An old friend, aged 71, said, "It is only when people are 60, 70, 80 years old that they begin to draw lessons from what they have seen, that they begin to understand about God. God delivers his light on the eve of death. . . . I regret not having known more about God while I was raising children. I worked like a slave, like a fool, dumbly, blindly, thinking I had to. But I didn't. Children will grow up whether we are dead or alive. My father was brought up by aunts in Tudanca. I myself was not bringing up my children, in any case. It was the hand of God." Often the older women become overwhelmingly preoccupied with the achievement of salvation.

Such women, sometimes nicknamed "beatonas" or "beatas" after a kind of lay nun common in most parts of Spain in the sixteenth and seventeenth centuries, are objects of ridicule among many men, even among some of the younger priests. The women are generally wronged in the process. One of the more sympathetic of the priests expressed a finer understanding: "They are very often women with a great deal of charity, with incredibly scrupulous attention to the smallest detail of their lives—whether or not they should have done this or that. Their faults might be due to their lack of education or their age, but they are often people who live the presence of God."

The ones I know have become devotional, even theological virtuosi. They are assiduous not only at public church functions, but also in the reading of prayer books, the performance of novenas (since they are many of them too old to get out to the chapels), and private meditations. Some of them make virtually total effort to live up to what they think a Christian should be. And in the process they have a good time. For the woman cited above, religion is no morose, waiting thing. It is quickened with job and discovery. Reading the *Guide for Sinners* of Fray Luis de Granada, her eyes light up with pleasure. For her and some of her friends, religion is poetry, drama, mathematics. The mysteries of the Church are for her a series of splendid jewels, and an endless one. Old age provides an opportunity to work out the logic of ceremonies, the relative merits of different devotions. Religion becomes an enriching, fertile preoccupation. The older women become, in their own way, mistresses of a vast body of arcane lore and tradition, philosophers and technicians of the sacred, consultants to their daughters and granddaughters, to whom they pass on their personal patrons and their techniques for contacting and consulting with God.

The changing states of feminine purity are evident from clothing colors. At first communion, age seven, the children wear white. Until this age they are considered to be pure and blameless. The woman wears white again at her wedding. She is still pure, if less so, for in the meantime she has undergone puberty, her entrance into the working world, and a certain amount of unavoidable pollution. After marriage she never wears all white again. The colors she can wear are muted. As she gets older they become darker and darker, partaking less and less of white. When she is widowed she abandons the last remnants of light; she wears black. The black is initially for mourning. I see it also as a statement of utter humility before God, a kind of uniform of abasement not unlike the cassock of the priest, a statement of intentional purity, as opposed to the intrinsic purity of childhood. Generally the frequency with which she receives communion increases. This is the time for purification. The larger cycle is completed: cleanliness in childhood, pollution in the active life, purification in old age. At death she is wrapped in a shroud of white, in testimony of her regained purity.

The cycle is less marked for men, and its reflection in dress is more subtle. This is because men have little contact with cycles of purification. Again, the cultural dichotomy between men and women is enormous. The women not only bear the weight of religion on their shoulders, but by and large they bear the sins of the entire culture.

The delineation of cycles of purification tells much about what the villagers consider more sacred, what they consider more profane. This sacred-profane distinction is not one artificially imposed on the culture by anthropologists; it is made explicitly time and again by the villagers as they sort out *cosa sagrada* and *cosa profana*. An additional distinction of this sort can be inferred from the examination of the purification cycle of emigrants. In some ways it seems to parallel the cyclical return to the shrine of the adult women. It points to a notion of the profanity of the world outside and the relative sacredness of the world within (the valley).

José adumbrated this syndrome when he pointed out that his devotion to the Virgin of Las Nieves might be stronger precisely because he had been away from the village. Time and time again, all across Spain, I was told by priests of the intense devotion of emigrants to the village shrine image, especially Mary, how at times emigrants would barely pause at home, particularly after returning from many years abroad, before going up to the shrine. I myself have accompanied returned emigrants on such visits.

Doubtless the major cause for this devout attention of emigrants (usually males) to the shrine image is the sense they have of Mary's personal protection over them in their odyssey. It has been traditional in many parts of Spain for men to pay a visit to the local shrine to commend themselves to the special protection of their personal patrons when they leave for military and on other occasions when a trip away from the valley was seen as fraught with dangers. It would be only natural then, upon one's safe return, to give thanks for the protection and fulfill the promises made during the time away. Certainly there is also the sense that while the village and one's family cannot accompany one overseas or far away, the Virgin is a family member who can; in a situation of aloneness, as in the many cases cited before (illness, widowhood, people with very private problems, priests), people draw closer to their God.

But sometimes emigrant devotion also seems to be a case of pollution and cleansing. Village ethics do not apply to the outside world, and it is clear from stories and songs that the emigrant is expected to be hard and ruthless in amassing a fortune to bring back to the village. The shrine, in such circumstances, becomes a cleansing station where a ritual purification (a kind of debriefing) takes place. The statement of a policeman in the city of Oviedo who comes every year to La Luz could apply equally well to many Indianos, "Every day in my work I have to involve myself in all kinds of dirty business *(sucie-dades)*, but every time I come here I feel I am washed clean."

For the Indianos the New World is a kind of free-fire zone. When they return to the village to marry local girls, they return to the Virgin at the shrine. The shrine is a redemption center that permits reentry into the village society on the old ethical terms; it is a washing station, a place where norms of the New World—of Mexico City or Guatemala City: the pursuit of gain—are shed and the ideals of the village—mutual aid, discreet living, chastity, and conformity—are reassumed.

Although the time elapsed is often shorter, and there is less of a cleansing action on reentry, the villagers who leave home to work elsewhere have the same double standards of morality that the Indianos have. The set of rules that apply to the village, to the tribe, do not apply outside. Outside it is every man for himself. "Well, you know what it is like in the city. One thinks only of how to get two pesetas out of one," so a shepherd replied when I asked him whether men from the village attend mass when they work (as he had worked) in the city. Another elderly man gave me a rambling disquisition one night on the subject as we brought the cows in:

> This world is made up of all kinds, with all sorts of different assignments. In this world there are very good people, good people, regular people, bad people, and very, very bad people. I have been all over Spain, in Navarre, two years working in a store in Cádiz, in Salamanca, Madrid, and in Burgos. I have seen very good people and very bad people, people who smile at you and take every penny you possess.

> In Tudanca are there very bad people?

> Here no. There are no very bad people.

> In Cabuérniga, in Liébana?

> There may be the odd one, but very bad, no.

> Then where are the very bad people?

> The very bad are in the cities. In the cities there are all kinds. In the villages such people would stand out too much—the villages are too small. So they go off to the cities.

> Will many people in Tudanca go to hell?

> No. Purgatory? Yes. They tell us that God pardons a great deal.

Because the city is an amoral place, all rules for behavior are off. Once the villagers, especially the men, start talking about their experiences away from home, their anecdotes consist of one case after another of how they were done in or how they did others in. The main virtue away from home is seen as craftiness (being *listo*), while within the village it is seen as a vice. Within the village openness and honesty are the most highly regarded virtues. The attitude towards the outside world in these situations is similar to that of the picaresque novels like *Lazarillo de Tormes*.

The active life outside the village, at least as recounted afterward, rests on the principle that human nature, if left to itself, has self-interest as the mainspring of its actions. However, in the village there are other assumptions about human nature—that people will to some extent share with you and work with you, for, as pointed out in the first section, the entire village is to some extent a communal enterprise that depends on joint effort and mutual aid.

While these are stances, they bear a healthy relation to fact. In large-scale societies people do have to look out for themselves. This is partly because as my old friend intimated the bad people of the villages have all been drained off to the cities. That draining of sinners, dropouts, and nonconformists takes place is undeniable. Catholic countries, especially in the current era, are involved in massive sorting operations by personality. Divorce, for instance, is illegal in

Spain: therefore desertion is not uncommon. In every village of 60 households there are one or two cases of women who have been left with children by men who have then disappeared into Bilbao, Barcelona, or Madrid. Similarly there are stories of women who have "gone bad" and taken up residence in cities. These processes, and others like them, are inevitable, given the lack of privacy and the oppressive morality of the village and its Church. From the point of view of the villagers the city is, among other things, a kind of sink for rejects.

In contrast the village (perhaps in part because of the evacuation of deviants) is a sort of utopia. My old friend, pointing out that there are some very bad people in the world, said it in a way that would indicate that this was not a normal observation, that anyone growing up in the village would think that people were, by and large, nice. This nuance was reinforced by his supporting statement that he had been all over Spain, the Spain he referred to being the cities he has worked in. Implicitly he was constrasting Tudanca and the world outside.

Because of the cycle of purification, emigrant men seem to develop a tenderness toward religion that the men who stay in the village do not. The emigrants are often pillars of the church and contribute funds for building repairs. The sacred-profane distinction that their particular cycle points to, that of valley and outside world, however, is made by all villagers who have had the chance to explore beyond the valley and even by those who have remained behind since the notion has become part of the local folklore.

The idea of the indecentness or unethicality of life in the outside world was reinforced in earlier days by the nobility and the clergy for political reasons. Some of the present-day attitudes may still bear the marks of these labors. This corner of the province of Santander was significantly more conservative and more devout than most of the rest of Spain, or even the neighboring districts, in the past two centuries.[58] One result of such a situation, one in which these valleys were in a sense backwaters, was the people's condemnation of the more liberal and modern world away from the villages as corrupt. This appears to have been done especially by the clergy and by the hidalgos who wanted to maintain their political influence. José María de Pereda's novel about Tudanca, *Peñas Arriba*, written in 1880, is an extraordinary case in point. Pereda's heroes, and through them Pereda himself, see the valley as a bastion of the finest traditional principles, threatened by vaguely defined "ideas" from the outside. The protagonist is a citified socialite from Madrid and the book follows his conversion to the more wholesome ways of the valley. In fact he takes over his uncle's self-appointed task of preserving, by the force of goodwill and the judicious expenditure of the family fortune, the political innocence of the people. The villagers in the novel are treated as idiosyncratic, happy, untutored children who should not be exposed to the evil of the modern world. The informing political ideology here sees the "progress" of modernization as the advance of evil.

At several points in Pereda's novel the evaluation of the villagers as closer to God than the protagonist, because of his time in the "world," is explicit. The novel and its author should clue us to the fact that this use of the corruption or inhospitality of the outside world is frequently encouraged by those in power in order to maintain the cohesion of a community and their influence in it. Power in isolated communities accrues to those who are the brokers with the outside world.

Since 1936 the rest of Spain, at least officially, has largely shared the political and social attitudes of the valley's elite. Hence for political reasons, at least, these has been no reason to maintain the devaluation of the outside. For whatever reason, many of the younger people and some of the older ones do not now share the attitude. Indeed, we have already pointed out the extent to which the villagers have been won over by the attractions of urban culture.

As previously pointed out, these different cycles themselves are internalized reenactments of the Christian story as presented by the Church. The Garden of Eden, the Fall from Grace, the life of toil, and eventually, the redemption have penetrated to the roots of the emotions and self-images of some of the people in the culture. On others it has had little direct effect. Many of the men are in this latter category. Its effects on them are indirect, coming to them through the behavior toward the men of the women and the priests who are actively engaged in working through the cycles. Although in the past men were reputedly more devout, my guess is that there has always been on their part a certain detachment from the pollution-purification process.

There is the ultimate cycle which adult males do, indeed, actively participate in. That is the sacramental cycle of birth and death. In this cycle one is conceived in sin, for any initial cleanliness was lost by Adam and Eve and there is no getting around that. All life is impure under these circumstances; only a good death and perhaps a certain time in purgatory will ensure purification and entrance into heaven. Most of the men are not so sure now that even this formula is a real one, but they are sufficiently unsure that they are very careful of their dying. The masses left for the souls of the dead by the dying in Obeso from 1615 to 1815 leave no doubt as to the past. The men were just as careful for the welfare of their souls as the women. And the testimony of priests in the valley (and that of Miguel, above) bear out this for the present day. This is one ritual of purification in which the entire culture participates, in a sense the climax of the introduction to moral knowledge implicit in the first communion.

In the process of purification and redemption are used most of the devotions listed in the previous section, and most certainly the shrine images, especially of the Virgin, for these are stations for redemption not only of individuals but also at regular intervals of the entire collectivity (whether it be village, vale, province, or nation). The use of the shrine image as the redemptive or purification agency is operative on a critical as well as calendrical basis. The private devotions are also agencies of purification.

Indeed, even some of the promises, those involving public penance of one kind or another, partake of ceremonies of purification. At first glance it might seem that promises were strictly instrumental, but from my experience and observation it has often seemed that they partook in their most public manifestations of a sense of guilt: the self-imposed obligations were undertaken not merely to repay God for some favor he had granted, but also because either the person was unworthy of having the favor granted in the first place or the unhappy situation that called forth the promise might itself have been a punishment from God that one was called upon to suffer. In the latter case the mortification involved in the promise could be seen as a suggested substitution for a divinely ordained misfortune. The fact that the divine saw fit to ameliorate the misfortune would then be evidence that the substitution suggested in the promise had been agreed to. Indeed, most of the self-imposed penances—reparation money, mortification, going barefoot, the wearing of habits—were imposed by the Church as penalties in the middle ages. This would qualify them as appropriate substitutions for divine punishments. This interpretation is supported by the older doctrines of the Church that ascribed most natural disasters to divine wrath. For instance, in the revised Consititutions of the Diocese of Palencia (1681), which I found in the church in San Mamés of Polaciones, storms are said to be evidence of divine wrath:

> In the time of tempests, thunder, lightning, and storm clouds, the Lord God shows that he is angry with sinners for their faults and sins, and so every Christian should approach his Divine Majesty to please him to relax his wrath and exhibit mercy: but in particular the parish priests and the rest of the Clergy should do this, going to the churches to pray, and utilizing the formulae and exorcisms that the Church ordains to calm the said tempests and storms.[59]

Another example from the same statutes regards the precept prohibiting work on Sundays:

> And now in the notes that have been presented to us by the Clergy we have recognized the same and even greater lack of observance of this precept in all of the places of this our Diocese, and not with little trouble it happens that God our Lord, indignant at this irreverence and neglect of his cult and of the Saints, sends us continually punishments of sicknesses, sudden deaths and violent, and most of the years the harvests fail, even when they were expected to be abundant, sometimes with the calamity of the locusts that have leveled the fields, sometimes with the lack of rains, and other notorious problems.[60]

In these cases God is seen not merely as a passive respondent who may answer when called, but as an active judge who dispenses scourges and afflictions in this world as punishments for sinful acts or states of being. There remains a residual belief in this sort of divine activity, which adds a note of appeasement and atonement to the dominant ones of petition and gratitude involved in public penance and promises.

In contrast, these are many communications with God, usually by way of intermediary divinities (Mary, the saints, souls in purgatory), that partake little if at all of the sense of pollution or even of the quest for redemption. I am thinking of the more practical requests and pledges. Many men, for instance, those who have evaded in one fashion or another submission to the Church's notions of sin, will make pledges of money or goods for the health of an ailing animal to, say, St. Anthony of Padua, St. Anthony the Abbot, or to the souls in purgatory. Men's promises to the shrines and chapels seem to be fewer; the women make promises for the entire family, but many of the promises of the women are very nearly purely instrumental also. Cecilia is a good case of this. In such a situation the divine intermediary is little more than a power broker who must be repaid when it has been helpful without further ethical implications. The divine intermediaries or sources of pure power in these cases are more likely to be the shrine images and the church images than the generalized devotions; most of all they are likely to be the *divus loci*, the active patron that is particularly associated with a particular place.

While involvement in purification and redemption involves also an intimate engagement with the divine—necessarily involves belief in the deepest sense— participation in the scheme of favors and promises does not necessarily involve either an emotional engagement or even a firm belief in divine efficacy or power. In the first place it is possible to regard the promise as a simple transaction, no more emotional than a purchase at a store. Secondly, participation in the scheme need not call forth an absolute commitment of belief, because the promise system is essentially based upon probability. Promises are not made or contributions given except in situations in which the outcome is uncertain. There is usually the possibility that if merely left to chance a situation could have a favorable outcome by itself. If the outcome is indeed favorable, then only in most unusual circumstances can it be regarded as absolute proof of divine intervention (in which case it is called a miracle). The promise is fulfilled on the chance that such intervention took place, without necessarily having to make a commitment of belief that it did for sure. If a minor sum of money or goods, or an amount of effort is at stake, this argument would go: What harm will it do to throw whatever weight can be mustered into the balance, if any. At worst a duro or a mass has been lost. And in the case of most promises, if payment is to be made C.O.D. then nothing at all is lost if the favor was not granted.

In most cases belief, as such, is not an issue. A man or woman does not sit down and think, "Do I or do I not believe today." And, in fact, having little or no occasion to formulate beliefs verbally, the beliefs themselves lie suspended between poles of commitment and scepticism, fluctuating and ambiguous. More important than talk about belief, which is indeed a problem in a plural society but not in a fairly uniform rural Catholic culture is attention to emotional commitment and act.

The two types of relationships with the divine—the one concerned with purification, the other with practical aid—are of course by no means exclusive, and most of the persons who have one relationship have some measure of the other. But they are useful to distinguish because I know well persons whose relationships with the divine fall almost entirely into one or the other category. As I said, generally speaking the women fall more into the former, the men into the latter. Core families are more concerned with purification and redemption, peripheral families with instrumental aid. Finally, making the distinction between widows and on the one hand, older married women, and on the other, younger married women, the former are more concerned with sin and redemption, perhaps because of their place in the lifelong cycle of purification. The latter, in their relations with the divine, are more keyed toward obtaining favors, particularly for their children. It seems likely that historically, the more instrumental, pure power aspect of the religion developed first, and the shift in emphasis toward salvation is a later addition that gained in effectiveness only from the late middle ages on. The struggle between the two ways of viewing relations with the divine is a constant one, the priest in recent centuries emphasizing the Christian power of redemption, the villagers having their own uses for their gods.

The different ways of communicating with the divine and making bargains with the divine are related to the ways that people arrange transactions with each other. Secular transactions of exchange can be arranged on a continuum from strict reciprocity to a kind of family communism.[61]

1. From the local point of view, the most primitive form of exchange is the cash nexus. This relationship holds when a villager buys something or sells something at fairs or anywhere outside the village. It is a way of dealing in those areas of the world where people are motivated purely by self-interest. It is a simple relation of immediate exchange and is perhaps the root metaphor of the entire society. Much religious discourse is in terms of money and payment. The equivalent to the immediate exchange relation in human-divine affairs is the promise that involves an immediate payment, say a duro in the souls in purgatory box.

2. A modification of this is the delayed payment, at the end of the month or the season, that is practiced among villagers with cash transactions or between

villagers and their storekeepers. Most promises (like the majority of any villager's secular transactions) are of this nature, to be repaid at the Saint's day, at the shrine.

3. The next step up, practiced among noncommercial people who have a rough acquaintance, is gift exchanging. An approximate equivalence, if not overpayment, is expected. Indeed, friendship is measured by the degree of unspokeness, the degree of generality of the exchanges. A couple of incidents will make the norms of this situation clear.

An Indiano explained how he had obtained a set of fine cow bells. Four of them he purchased in Cabezón de la Sal, and the fifth he offered to buy from a shepherd. The shepherd said, "No, I won't let you buy it. But I will give it to you." The Indiano later gave him a very fine lamb that had very good wool and was a good breeder. He told the incident in praise of the shepherd's character, as if the exchange transaction was on a higher ethical level than a cash transaction.

Another shepherd, who happens to work for the same Indiano, told another story, this time about a similar situation that miscarried. One day after the market in Puentenansa he found a stray calf. He reported it to the Presidente of his village, and a few days later the calf was claimed by a cattle dealer who turned up from Liébana. In the meantime the shepherd had been feeding the calf with his own hay. His friends counseled him to ask the Lebeniego for compensation, but he refused, considering this an indignity, and he provided the Lebeniego with a helter to secure the calf in his truck. As in all situations of social intercourse, the formalities were observed, in this case going to the bar for a drink; still the man made no mention of compensation to the shepherd. Eventually he drove off with the calf. The shepherd kept his counsel, expecting that perhaps over time the cattle merchant would do him a favor. But a couple of months passed without any approach from the merchant, although he came to the Puentenansa market. Finally, one day when the merchant was in the midst of a bargaining session with another man and was being unreasonably intransigent over a price, the shepherd exploded and called him down. "You (Usted) are not a man. You have no shame. I cared for your calf and gave you a halter, and you drove off without so much as a word!"

The example shows that the system of exchange of gifts, when among those who are not well acquainted, is a fragile one. It is usually performed with dignity and honor, but the norms of reciprocity still underlie the transaction, and a fairly strict accountability still holds. This relationship is still a market relation. It is the kind of delayed reciprocity that the normal promise takes, for the normal promise to the divine does not involve repayment in any specific year. Just as men like the cattle merchant, are still held accountable for this kind of exchange, even though it is delayed, the mythology of the shrines includes examples of divine impatience and wrath when after a suitable interval the promise has not been fulfilled.

4. A still higher level of generalized transaction holds when an acquaintance is of long duration. Here the equivalencies are very much rougher, with little attention to measurement. The mutual aid may balance out in the end, but wider inequalities of obligation are tolerated. This relationship has a name. A person is said to be *de confianza* or *de casa*. It is generally accompanied, as the terms would indicated, by mutual openness of homes, including the right to eat and stay overnight unannounced. In a sense I provoked a certain number of these relationships by doing what would be considered very great favors for people, like spending two days helping them find lost animals or taking them on urgent trips with my scooter.

A clear statement of the relationship came one day in Tudanca. Ramón was missing a cow and came to see me about it. I took him to several villages looking for it in the rain over a period of two or three hours. After about an hour and a half he said, "We order you around like you were de casa, don't we? Well, you are *de casa. De toda confianza* you with us and we with you."

This relation is also described in the expression "as if we were family." It holds particularly between families in different parts of the province. The tie is used less frequently than it would be if both families were in the same village, but perhaps this way it is easier to maintain. One family I knew in Obeso was *de confianza* with a family high in the mountains of Campóo, where their cows graze every summer. this relationship served both families because the man in the mountains was a cattle buyer who often came to stay in the Obeso house at the time of the Puentenansa markets, and the Obeso man stayed at the mountain house when he went up to check on his cattle. The two of them never engaged in any commercial transactions that I know of, since that might have endangered the more valuable friendship relation.

5. It will be seen that the spectrum of relations successively approximates kin relations, which indeed is the logical pole of the series. In the first chapter it was noted that the two most important social structures for most persons in the valley were the village and the family. While both the village and the family to some extent share the characteristic of organizations of persons with a common goal, this feature is very much more pronounced in the family than it is in the village. The effect of this difference can be dramatic. For instance, members of the family work for the family; members of the village work for themselves. Only when the villagers contrast the village to the city do they emphasize it as a mutually aiding community, which indeed it is by comparison. Within the village the more impersonal norms of intravillage relations stand in contrast to the norms of intrafamily relations.

Self-interested activity for economic well being, we have seen, is treated as profane activity. Activities for the good of others or implying a trust in others are considered, by and large, as better. This accounts for the contrast, on the one hand, between village and outside world, on the other, between family and village. The modes of exchange for transactions between the valley and the rest of the country are predominately of a strict reciprocity. Social intercourse on the village and valley level is characterized by a diluted reciprocity, more diluted as people are better acquainted. But all such relations contrast with the authoritarian communism of the family. The family is a coordinated work unit, theoretically taking from each according to his ability and giving to each according to his needs, under the administration of the father. This distributive system has its authoritairan aspect mitigated by the presence of the wife and mother who can influence the father and who, because she is not the taskmaster, is free to develop affective relations with the other members of the unit.

An immediate equivalence to this distributive system can be seen in the roles of God and Mary and those of father and mother, collectors of grace and merit earned by the children, and distributors of graces and punishments as they are deserved. The affective, personal relations that characterize close friends and family relations of sibling, in-law cousin, and mother-child are found in the institution of personal patronage.

The villagers have before them this spectrum of relationships in which they daily participate. Surely it is not surprising that the entire spectrum is used also in regard to divine figures and that the contrast between a kind of mutualness within the family and the progressive reversion to relations based on reciprocity as one gets farther away from the family also characterizes the religious attitudes. This contrast virtually attains the status of a personality conflict within the individual. Within the village, among villagers, and even in transactions with strangers this is a lip service, a wistful, lingering attention paid to the notion that any particular exchange is not a market transaction but rather partakes of the more valued family relationship. Shopkeepers will intimate that it really does not matter to them whether one pays or not. Persons from whom one borrows tools will intimate that they do not really care when they get them back. Repayment of favors is disguised as spontaneous gesture. It is as if there were a kind of shame in the pure market transaction. Similarly there is a constant attempt at humanizing market transactions with God, paying more than one promised or generalizing the relationship with a shrine image so that it involves more communication than merely practical affairs.

Here is another of the sources and manifectations of the male-female dichotomy. The men, whose relationship to the divine is generally less affective and more instrumental, are more involved in the social network of villagers—in the peer group of other men that nightly meets at the cafe or that patrols the hinterland—than are the women, who are by their tasks more intimately involved in the social relations within the family. Time and time again the isolation of the man from the family unit has been evident, his distance from the children clear.

His attention is elsewhere, in the male society. Man and woman, then, in their relations with the divine, are duplicating the primary modes of the social relations that stem from the social units in which they chiefly operate.

It is clear from sermons and religious literature that at the parish level the priesthood (and this crosscuts current ideological divisions) has consistently attempted to use the family analogy as a model for divine-human relations, as opposed to the market analogy. This stance has entailed, in the Church as a whole, the modification of dogma to make the analogy better—the establishment of Mary as an important figure to correspond to the mother. Indeed, the priests have also consistently tried to widen the application of communal norms from the family to the village at large. This kind of pressure partly accounts for the universal refusal to accept unmitigated pure market relations at the village level. The Church has from its inception been opposed to instrumental exchange relations among humans; surely that was one of the essential messages of Christ. In its attempt to supplant these relations the Church has been thwarted by its own right hand, for as an institution it needed some kind of regular system of income, such as that provided by promises, to maintain itself. Tithes are contributions in the image of family contributions. But the shrine network is different. In a sense it stands in opposition to the parish network in that it encourages the kind of exchange doctrine that the parish priest has generally opposed.

Whether as cause, effect, or coincidence, the norms of transactions with the divine follow norms of transactions among humans. In fact, there seems to be a kind of equivalence between the different levels of intimacy with human figures and the different degrees of intimacy with divine figures. The personal patron is closest to a family figure. As in the case of Gerónima with the Sacred Heart, and José with Las Nieves, these devotions are conducted in a personal, uncalculating way. Gerónima, Angela, and Cecilia, however, all have relations with other divine figures, whether in the church or at different chapels, that are less intimate, involving simple transactions, delayed, fixed-time payments and delayed long-term promises. In other words, there are people for whom the range of transactions with the divine duplicates the range of transactions with humans. Similarly among men, whose relations with the divine are likely to be more formal and reciprocal, matching their social relations, there is one exceptional relation that occurs with regularity, the relation to Mary. This exception matches the one long-term relation of affection that they maintain in adulthood, the relation with their mothers.[62]

The parallels between person-to-person and divine-person relations are often explicitly drawn and used in the valley. Christian religious doctrine is a given, and the parishioners use whatever analogies seem useful to elucidate the doctrine that they are presented with. The Church itself has eased this process by providing suggestions ranging from royalty-subjects to judge-lawyer-petitioner to shepherd-sheep to patron-client to Lord-servant to parent-child and most recently to friend-friend. The people's explanations and understandings of doctrine will partly be a function of the repertoire of available secular relationships that they can use as explanatory analogies. They will emphasize those aspects of doctrine most analogous to the particular relationships they value. The analogies work the other way, also. The people may be influenced by the divine-human and divine-divine relationships postulated by the Church in their own day-to-day social relations.

One must distinguish the mechanics and norms of social intercourse, whose variety, already discussed, is replicated in relations with the divine, from the positions of status that people bring to social intercourse. In the course of their lives all villagers have dealings with people more powerful than they, beginning with their father and moving up to the priest, the lower ranks of the nobility (in the old days), and the valley professionals and government bureaucracy. They also have dealings with people of equal power: their siblings and peers. All but the most benighted deal with people of less power: less fortunate vecinos; the odd tramp; and most important, children. Even children are in a position of power over their animals, a tyranny which they often exercise with obvious delight.

But not all of these power relations are useful as analogies or models for their relations with the divine. In the brand of religion until recently predominant, divine-human relations held most analogies to relations on earth that included an imbalance of power and there was little or no place in the divine-human repertoire for peer relations. Likewise, in many of the social relationships where there is an imbalance of power, the imbalance is not absolute. This disqualifies these relationships as models for God-person relations, for its essence is the complete power of one over the other—if not immediate, then ultimate. Such power is held in the valley today by Franco and his bureaucrats, by judges, by fathers over sons, and by masters over animals. These then are the most useful analogies currently employed by the Church and by the people for God-person relations. The use of Franco is one substitute for the older analogy of the heavenly kingship. Similar analogies formerly in use were those of master-servant, owner-slave, and Lord-vassal.

All of these sources of power, with the possible exception of the father, are remote and unapproachable. It is almost a question of linguistics. Judges have a special kind of language, as does the bureaucracy that leads to Franco, and animals cannot speak the language of their masters. It is therefore necessary to use intermediaries when negotiating the various kinds of transactions that we have discussed above. In the case of Franco (or, more generally, bureaucracy "out there") persons go to literate friends or urban patron families who can put a word in for them. In the case of the courts one employs lawyers. In the case of the family one uses the intercession of one's mother or of one's father's brothers. Indeed it is with this layer of middle-men that the whole gamut of relationships and techniques discussed above becomes useful, depending on the length and intensity of one's tie with the intermediary. It is the middleman who is the "patron" in the secular patron-client system. He or she is the person or agency who has not absolute, but relatively more, power than oneself.

The three-tiered system of power, with intermediary translators coding and decoding messages from one system—say the oral valley community—to another—say the literate formal bureaucracies beyond the valley—corresponds in its essentials to the three-tiered system of relations with the divine. "It is as if Jesus Christ were Franco and the saints were his ministers. Instead of going to Franco people go to the ministers and use them as intercessors. They are closer to us" (. . . a carpenter in Puentenansa). "The saints are useful just as you would call on a friend or an uncle in Torrelavega if you had to get something done there" (. . . an antique dealer in Labarces).

There is a certain number of things that one's patrons on earth can do for one by themselves because of their education, wealth, or cosmopolitanism. In any case, since the ultimate sources of power are generally unapproachable, virtually all of one's attention centers on the patrons. They are, after all, the ones one must please. Similarly with the saints. It is clearly understood that the saints and Mary are not the same as God, but they are the ones whose attention is cultivated. They are the ones to whom the various secular techniques of exchange and intimacy are applied.

The saints referred to in the previous quotations are the chapel images and the generalized devotions enumerated in the second section. As well as Mary, Christ himself may be included in the category of saints. The man from Labarces had as his personal patron, for instance, the Infant Jesus of Prague. Although the Church may introduce such devotions, like the Sacred Heart of Jesus, as a way of humanizing and making more accessible the people's image of God, they become reified intermediaries like the other saints to God himself who remains, in the public eye, like the secular powers that be, too far away to be approached.

In the past 100 years the Church has turned away from the feudal, the monarchic, and the other political images because of their lack of universal applicability (*vide* France and the United States, to begin with). It has increasingly concentrated on the family as an inductive model for understanding divine relations and *vice versa* on divine relations as a model for the ideal family. Because the one intercessor of importance in the family is the mother, this has made irrelevent, by and large, the vast array of saints to the catechistic analogy. This may help account for the relative decline of devotion to the saints as evidenced by the falling off of saints' names, the collapse of saints' chapels, and the recession in prayers to saints that we examined in the previous chapter.

Be that as it may, the analogies of Mary to mother and God to father (and its unspoken corollary, people to children) is the major stock in trade of preachers in the valley today. Most sermons argue from the self-examination of one's own warm mother-child relation. The analogy works two ways: God is the son of Mary. That makes her a perfect intercessor, because what son refuses what his mother asks of him. Mary is the mother of us all: she will care for us as a mother does and listen to our pleas.

Ten of the fourteen villages in the valley have a strong village patron that is an image of the Virgin. In these cases, metaphorically speaking, she has the role of the mother (i.e., protectress) of the entire village. The village itself then becomes, metaphorically, the household of the villager. Consider the explanation a herdsman from Obeso gave for the special obligation Obeso people feel for their patron, Our Lady of the Lowlands:

> People generally have special devotion to their village patron. When there is the fiesta of Our Lady of August (= Lowlands) the entire village turns out. People feel proud. Their self-respect demands that the fiesta be a success. Similarly Cosío people turn out for their fiesta. This does not mean that there is a distinction between the devotions, between the images. Its like if you and I were sitting here, and you just told me what good friends we were, and then your brother needed you, then you would go to your brother. You choose a brother over a friend, that is normal.

Here the metaphorical links between the village patron and a family member and the village and the home are explicit.

With this analogy the symbolic role of Mary as village patron becomes a very important practical one. She is a significant factor in binding the villagers together as members of the same family. Just as the extended family stays together as long as the widowed matriarch is still alive and active, so the village stays together under the continued activity of a powerful image. This is how the very essence of the village as home can come to be bound up in a revered image.

This of course is a more exclusive form of the wider Christian doctrine of the brotherhood of all men under God the Father. The specific limitation in this case of the divinely delineated family to the village membership is precisely one of those factors at the root of a sacred-profane distinction made earlier—that of the village as more sacred and the world beyond as more profane, which permits unethical conduct when one leaves the local society. The intermediate regional shrines around the Nansa valley, drawing as they do on disparate, politically fragmented hinterlands, serve to cultivate little or no sense of community among their devotees.

What is the implication of this complex (village-home, Mary-mother) for the life cycle? A divine patron is generally not needed by children, as their parents fulfill most of these functions. The shift from dependence upon a earthly mother to that upon a heavenly mother generally does not come until some measure of financial independence from the family work unit has been achieved, i.e., until the earthly parents are no longer functioning for the individual as the administrators of labor and the distributors of goods.

American couples without religion—or more simply, without parent substitutes—living away from their parents tend to find in each other both parent and friend. Or at least there is more of a mix-up of these things in mobile, middle-class America than in the Spain of the village. When couples in America have children, then the roles of mutual parents become incompatible with actual parenthood, and it is often at this time that couples begin attending church again. In Spain this childless period does not occur: The man soon deserts his wife for an all-male peer group, and the wife is left alone with the children, her mother, or her heavenly mother. The older a woman gets, the less likely she is to have a living mother who fills all the needs that a heavenly mother can fill. And old people have nowhere to look for help but up.

What of the obverse effect—the effect of the divine-human relations on the secular relations of person to person? To begin with, the family: The analogizing and mutual modeling of divine figures on family and family on divine figures has gone on for so long that it would be ridiculous to say which, at present, has more effect on the other. Similarly, as both change—as both the family and the notion of the gods become more relaxed and less authoritarian—it is very hard and perhaps not very important to say which is the independent and which the dependent variable. In the older pattern of family living and religious precept the Church actively encouraged norms of family behavior which, coincidentally or not, corresponded to the image of divine-human relations. These norms, reproduced from the Claret handbook (1832, 1946) in Appendix I, specified that the father-child relation should be that of absolute authority governed by justice; that the mother-child relation should be one of affection and love, with obedience through love.[63] There is no mention of obligations to one's siblings or one's neighbors. As pointed out already, there were no divine-human models for this relationship, because all divine-human models were based upon authority.

These precepts are provided for the convenience of persons preparing for confessions so that they may confess *as a sin* their not having acted in accordance with these rules. In the catechism taught in the valley today it states that one of the blessings of Adam and Eve in the Garden was that they knew "the appropriate things for their station" (el conocimiento de las cosas conveniente a su estado). By implication this was lost after the original sin, and people began to violate the rules governing their station. In other words, one of the inherently sinful characteristics of humans, for which constant correction must be made, is that they do not know their place in society. By inference a certain structure of relations in society is divinely ordained and must be maintained by the Church.

For example the handbook details other kinds of divinely instituted obedience, for instance, that of servant to master. The similarity of these person-to-person rules to the contemporary concepts of divine attitudes toward humans is too complete to be coincidental. Clearly the Church took seriously the notion that God made man in his own image.

The relation of husband to wife, under these rules, demands special attention. The husband is told to have patience: the wife is told to be obedient. Here is a special case with no clear parallel in divine-human or divine-divine relations. (The theological formulation of the Church as the Bride of Christ is hardly useful as an analogical model.) But there are other ways in which the religion and its imagery serves to reinforce "desirable" patterns of relations between spouses. I will let several men from the valley speak on this point for themselves. As I was going to the shrine at Bielba for the fiesta I talked with a local man about religion. He said, "There's one good thing about the Church: It civilizes people. Without it we would be acting like beasts; it is a brake on society."

Four days later, with a group of men in a bar at a cattle fair, the mayor of the village spoke to me, apropos of my Protestantism, "The worst thing about Protestantism is divorce. Catholicism is very convenient as a religion. It keeps families together—the sons obey their fathers and the wife keeps in her place." Soon after, the role of religion in marriage was stated most explicitly by a man from the valley visiting the shrine of Covadonga with me: "Men want to marry virgins, or at least be the one the girl fell to." I said that this was no longer a preoccupation in much of the United States. He replied, "It is very important here. Only if the woman doesn't know anything else will she think her husband is the best fucker around. If she has played around before marriage, then she is more likely to make her husband a cuckold, or even run off. The woman's place is in the home. She must be a solid wife and a solid mother." He looked over the people around the shrine. "You know, the Virgin Mary as a example is very influential upon women, keeps them in line. Not that the Virgin was fecund, but that she was a good mother, and pure."

This pragmatic, conscious use of religion and religious models to maintain male dominance over their wives and over their children (and of authorities over the men) may be more or less characteristic of the men in the valley. But the truth of the effect of the models is undeniable. The participation (one is tempted to say the entrapment) of women in the cycles of purification may cause their masters some distress over time, but it ultimately serves their masters' purposes and protects their masters' honor.

In this culture, then, there has been a reciprocal modeling that has taken the following form:

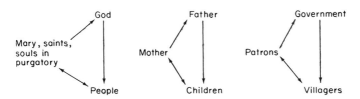

These are the ways that hierarchical human systems are models for and are modeled upon the hierarchical divine-human system.

An argument might be made that the family or political organization is the source for the relationship with the divine. Such an argument would run that the hierarchical family relation is the most expeditious form to exploit the land or that a hierarchical national organization is the most efficient way to face a military threat. In other words, people might say that the family and the nation are naturally hierarchical in organization. This argument might then reduce the divine relation to a replica of these social forms. On the other hand it is certain that the forms of divine-human relations have deeply colored and at the very least reinforced some of the hierarchical modes of human social relations. It is impossible and perhaps unimportant to say which pattern came first.

A similarity has also been shown between the relations of adults of equal status and the instrumental relations of adults to divine figures, both in the range and the variety of relations. Here it would seem to be a clear case of human modes of relations influencing ways of treating with the divine. There is no place in Catholic theology for the various kinds of bargaining with the divine that are common in present Catholicism. These are clearly derived from modes of human exchange.

The ties to the system of social relations are not only very close to the patterns of human relations but are also closely related to the *occasions* of human relations. In the life of an individual, the divine provides substitutes for missing or inadequate family relations. The more alone a person is the more likely that person is to turn to the divine and establish an ongoing association. Divine relations of an affective nature tend to fill in the gaps of human relations. Behind this view is the concept that one purpose of all organized society is the protection of people from aloneness. In situations of aloneness, religion fosters the creation of what might be called a second self, and sponsors the ability of a person to hold a dialogue within himself or herself. It could be put the other way around however, that a person's most natural companion is God and that in the course of human affairs relations with other people can get in the way of companionship with the divine. That is the principle behind monastic life: it is the way many priests and some of the older village women would explain it.

Be that as it may, both points of view might coincide in the principle that persons, if they are to survive psychologically, must develop a sense of personal value, a sense of self. This sense of self may be built up through interactions with others, but if human others are lacking, it may be built up through interaction with God or other divine figures. In this valley in general the women have more intense relations with the divine; the men have more intense relations with each other. These seem to be two alternate modes of establishing and maintaining individual values and identities as people.

In the search for the sources of the common modes of divine-human relations, then, one of the sources is the social relations that the villagers experience in their daily lives. Another must perforce be the theology, mythology, and traditions of the Church whose rituals so permeate their culture. On the one hand there are indigenous patterns of social organization (leaving as unanswerable the question of how informed they might have been at the origin by theological notions), and on the other there is cultural information brought from the outside by priests. It is to this cultural information that one must turn for the purificative or salvationary mode of relating to the divine and for notions of what is pure and impure.

The whole mode of purification and repurification throughout life, with the ultimate alternatives of different destinations after death, is taught and maintained by the priests through the different sacraments of the Church. But many of the local notions of what purity, impurity, and profanity consist are not explicitly taught by the Church. In fact some of them, like the profanity of the active life, the impurity of menstruation, child-bearing, and sexual activity, are if anything contradicted by the Church's teachings. Yet the Church consists of much more than what it provides as doctrine. It was seen, for instance, how the ritual of churching contradicted the expressed purpose of churching. Rituals can be a way of expressing a state of relationship between persons and God. Likewise in myth or religious history, as learned in the schools and in catechism, there may be implicit lessons.
R. W. B. Lewis has put this very well:

> While the vision may be formulated in the orderly language of rational thought, it also finds its form in a recurring pattern of images—ways of seeing and sensing experience—and in a certain habitual story, an assumed dramatic design for the representative life. (The Passion of Christ, for example, is the story behind much Christian argumentation.) The imagery and the story give direction and impetus to the intellectual debate itself; and they may sometimes be detected, hidden within the argument, charging the rational terms with unaccustomed energy. But the debate in turn can contribute to the shaping of the story.[64]

Instead of intellectual debate and argument, we have been studying ways of living, but the kind of informing of life and "charging" of ways of living "with unaccustomed energy" surely applies. The story of the Garden of Eden, the Passion of Christ, and the Last Supper seem to be latent configurations in the Catholic culture of these villages. They lie within the people and affect the way they act.

Indeed, the Passion and Resurrection of Christ and the Last Supper are not even latent configurations but analogies explicitly drawn by the clergy to the sacrament of communion and the possibility of redemption after death. But the story of the Garden of Eden is less commonly mentioned in the village religion. It is taught in catechism as the source of original sin and the cause of mankind's exposure to the passions, to sorrow, and to death. But the verses in Genesis are more specific: As penalties for eating fruit of the knowledge of good and evil the Lord God condemns Eve to the pain of childbirth and the rule of her husband, and Adam to toil in the fields. Here, as Lewis would put it, is "a dramatic design for a representative life." The subjection of woman to man, the relative impurity of woman because of her role in the original crime, and the unsacredness of toil are all programmed in this story. Models for redemption are made clear by the Church in the figures of Christ and Mary: patience for men, purity for women. But the origin and nature of the impurities inherent in the human condition are left veiled. They are there to see, however, in the rituals of the Church, in the structure of the Church, and behind these, in the informing story of the expulsion from the Garden.

All of this is not to say that all these notions and taboos are necessarily exclusively imports. They characterize numerous primitive, peasant and modern societies. Rather, it could be said that the way they are understood and enforced in the valley today is through a Catholic mythology and religious ritual. The pattern of culture itself has clearly affected the configuration of the Christian story. This is what has happened throughout Mediterranean Catholicism over time in the case of the progressive increase of importance of first the saints, then Mary in Catholic theology, to adjust the religion better to the structure of the society. This has occasioned, in a sense, a continued rewriting of the Gospels and continual reinterpretations of its meaning.

In summary it may be said that behind those modes of approach that involve treating or treating with the divine as if the divine were operating with human rules—what we have termed the instrumental and affective modes—are human patterns of behavior of long standing that continue to this day in the valley. Supporting the modes of approach to the divine that see the divine as a state of being are theological or mythological models brought into the culture by Christianity.

XI. Conclusions: Three World Views

Until the turn of the last century, and in some villages until the 1930's, the parish priests were local boys without a great deal of education. They often kept cows themselves and sometimes wore regular work clothes. The Episcopal Visits recorded in the parish account books bristle with written admonitions for priests to wear their cassocks and collars at all times. Such priests, because of their membership in the kin network and their insertion in the culture, were usually not terribly innovative. Their emphasis was on a proper fulfillment of the obligations of the community and the individual to God and the maintenance of the devotions and brotherhoods launched by the missionaries.

By the turn of the century, more adequately trained priests from elsewhere in the diocese were assigned to the valley; they brought with them, as we have seen, successive waves of new devotions. Their main struggle was to ethicalize the religion, to change it from an instrumental set of outward observances into a deepened, more inward spiritual life. The relationship they emphasized was that of the individual and God, and they encouraged the personalization of the relationship. This was a movement all over Europe, and the sentimental style of lithographs, the statues of saints with the baby Jesus, sweetness, and light characterized the whole period.

As the century progressed and the Church became visibly threatened by the force of reason and revolution, a toughness developed, a note of belligerence. But all through these different tones and colors, the philosophical groundwork of the Council of Trent remained. This groundwork, which on the level of the individual had as its chief effect the provocation and maintenance of cycles of purification and redemption, should be seen as overlaid on a far older set of principles and activities. These are the activities that deal with the very land itself, that see the landscape as brimming with meaning and coherence. While the generalized devotions, designed as aids to salvation, belong to the post-Tridentine period, the old chapels belong to the pre-Tridentine (one is tempted to say the prehistoric) period of valley life.

The two modes of devotion exist side by side; indeed they interpenetrate and make mutual adjustments. With few exceptions the priests tolerated the elaborate if informal system of instrumental promises and concomitant mortifications aimed at the chapel images. And some of the people acceded to the clerical pressure for purification and redemption. The chapel images became agencies for redemption as well as practical protection. Conversely some of the generalized devotions presented by the priests and the orders as auxiliaries to salvation, such as Our Lady of Mt. Carmel, the souls in purgatory, and the Sacred Heart of Jesus, were adapted by the people for use as patrons and shrine images, utilized for practical aims.

This interpenetration has slightly blurred what is an essential difference, both historical and philosophical, between shrine images and generalized devotions, instrumental and purificative religion. The older set of divinities are not as much intercessors with God as they are intercessors with nature, for they are located in specific places in the village landscape, from which they can be moved only under certain rigorous ritual conditions. They mark off boundaries between village and village and boundaries between cultivated and uncultivated land. Throughout Spain they mark critical points in the ecosystem—contact points with other worlds. Mountain peaks, springs, and caves seem to be contact points with the worlds below and above; boundary shrines with other earthly worlds. With their periodic devotions at these sites it would seem that the villagers were at once confirming the boundaries of their world, assuring the continuity of the annual cycle of seasons, and attempting, through propitiation or the use of promises, to gain some control or some influence upon the entry of foreign material or foreign power into their world. Since they themselves are not capable of fully regulating their environment, there must be other powers beyond who are capable, and the chapels are located at the most logical transaction points with these powers beyond. Individual promises and individual regular devotions served the same ends for the individual as the village ceremonies did for the village—to influence the course of crises and ensure the normal unfolding of the life process. The religion brought in by the priests, penetrating to most of the people in this region only after the Counterreformation, emphasized the Christian message of salvation and a series of purificative actions and ethical principles useful to that end. Communication with other worlds was to be mediated by the Church and its ministers.

With the breakdown of community boundaries through mobility and the media, with the industrialization of Europe, the rise in standard of living, the circulation of alternatives to the Catholic life, and by way of one brand of Catholic response to these events, the Catholic Action movement and the Second Vatican Council, a new model for communication with other worlds has developed in the valley. It has been brought in by radio, television, students, returning emigrants, but above all by the young emissaries of the Council, the younger priests. Often hostile to the older priests, certainly opposed to many of their sociopolitical assumptions and devotional practices, and critical of their nonaggression pact with the local pantheism, the younger priests are having a decided impact on the valley. The impact can be measured by the incredible reaction of the village and its gods—through a series of apparitions of the Virgin Mary in San Sebastian in the early 1960's. The impact can also be seen in the partial conversion of the youth to the ideas of the young priests and in the revalorization of democratic, mutualistic relations on all levels.

The new doctrine of these young priests virtually renders divine images irrelevant, or at least radically challenges their usefulness. As they are often products of village Catholicism themselves, many younger priests maintain a reverence and affection for Mary. But they do not regard either Mary or the saints as essential intermediaries with God. Much less do they regard the geographically located shrine images as special points of access to the powers beyond. Instead they have found a new intermediary between God and man: man himself.

The priorities of God-person, person-person relations for the older priests could be diagrammed as follows:

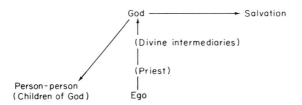

The use of intermediaries helps the person to form a bond with God. As a consequence of that bond he or she is then a member of the family of mankind and is also on the road to salvation. The prime emphasis is upon the relation of God and person and the eventual salvation of the individual.

The priorities of the villagers, *grosso modo*, are as follows:

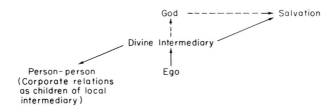

The intermediaries are useful in, of and for themselves. Salvation comes through a combination of their direct action and their intervention. Again, there is a kind of byproduct of community, based on the effect of a specific local intermediary as corporate parent or protector.

It will be seen how these two systems, sharing the same basic configuration, could live with each other. However, neither can live with the priorities of the younger priests:

EGO————▶ PERSON-PERSON————▶ GOD ------▶SALVATION
                (Brotherhood in Christ)

In this third system other people are the intermediaries on the path to God. There are no divine intermediaries. The bottom is knocked out of the system of promises. Salvation becomes the byproduct of the good life, not its goal. The priorities are completely reversed. Now in San Sebastian the money in the alms box that used to go for masses to the souls in purgatory (from promises, for the salvation both of the souls themselves and the villagers' temporal and eternal needs) goes to the diocesan charity fund. Living dependents replace dead ones; living intermediaries replace dead ones, whether souls in purgatory or saints.

This new system invalidates to a degree all the old distinctions between the sacred and the profane. The unvalued, uncontrolled relationships of peers, not even mentioned as religious obligations in the Claret handbook, are revalorized. The young priests associate with and encourage the teen groups, for instance. These were the most unsacred associations in the old system. In fact, under the old system, as we saw in Tudanca, the teen groups were a temporary counter culture which said mock versions of the Lord's prayer and mocked all authority by their actions. Because the mocedad was a mutualistic, democratic organization it had no place in the authoritarian symbol system, a system that was reinforced by, if not derived from, the two systems of priorities diagrammed above for older priests and villagers, in which the vertical God-person (or divine intermediary-person) relation sets the pattern for all religiously valued relationships.

The young priests revalue also the relationship of men among men. We have seen how, because the peer group was not valued in the old religion, this greatly contributed to the men's dismissal of the old religion as irrelevant. Now, especially in the cities, men are finding the religion more relevant. This has not yet happened in many of the villages because of the traditional hostility of men to priests and because the young priests have no idea how to approach the older men. Similarly, because person-person relations are most highly valued, certain kinds of mutual, community activities are preferred by the younger priests to the old habits of meditation and contemplation.

Because salvation is no longer a direct goal (a young schoolteacher told me she thought worrying about salvation was like loving a father for the money he was going to leave you), emphasis is no longer placed on purification, decontamination, and the domestication of instinct. Instead of an emphasis upon the suppression of the bad aspects of human nature, the young priests try to bring about an active transformation of self that permits one to do new things, instead of not doing wrong things.

Again, because priest-parishioner relations, like other relations in the society, are seen as mutualistic and nonauthoritarian, the priests shed their cassocks and work with people in the fields, or as bartenders at fiestas. In contrast, the older priests maintained a ritual purity, a distance from the population that enabled them to be effective negotiators with the divine. The younger priests say "we" instead of "I and you" in their sermons. Generally misunderstood by the villagers, they attempt to share as much of their lives with the people as they can, to break down the old isolation. (In fact, so far all that has resulted from their efforts is their conversion to the role of village-wide patron—their use as intermediaries between the village and outside bureaucracy that leaves them with fully as much power and authority in the village as their predecessors, if not more. In a sense, they have taken over the old role of benevolent noblemen.)

Naturally, the images of God have changed also. The younger priests encourage the concept of Christ as friend or model, although they do not find objectionable the notion of Mary as mother (that model still still stands). Two sermons preached by two types of priests in two separate villages on the same day show the shift in emphasis. The text assigned by the diocese for the sermon was the miracle of the fishes and Christ's words to Simon Peter, "You shall be a fisher of men." The moral of the text to be drawn was that each of us has an apostolic mission on earth. The older priest emphasized Christ's power in performing the miracle and drew a parallel in Christ's role with men, that Christ alone gives us the power to go out into the world. In contrast, the younger priest spoke of the ideas and example of Christ, not his power, in stimulating a Christian life. One emphasized divine help, the other divine inspiration.

Ultimately the position of the younger priests leads them to a denial of the doctrine of original sin and a revalorization of human nature, which they see partakes of Christ. Thus, the kind of human brotherhood they bear witness to is less a brotherhood of the children of God, as in former theology, and more a brotherhood in Christ himself, based on the sacrament of communion.

As a result of this new doctrine, certain things are lost. The impact on the villages is beginning to be felt, but not yet with full force. First of all, there is a philosophic loss. Under the old system, both the philosophy of the village and of the Church made it possible for people to perceive a world behind their world. Immanence was a reality. Virtually every shrine in the indigenous traditions and virtually every devotion brought in by the priests was started by revelation, by the appearance of a divine figure on the earth who demonstrated the usefulness of the devotions, who pointed out the ceremonies and rituals that were to be fulfilled, who in essence sealed sacred contracts upon which were made conditional the natural unfolding of life. This is the true message of most origin legends of shrines. Behind all events was meaning for those who had eyes to see and whose hearts were open. In the daily lives of the active villagers, these meanings were only fleetingly perceived. They were accorded a special status in the universe of truths—they were left in an anteroom of suspended disbelief. To this category of neither truth not falsehood applied the sacred condition of certain animals, like the mule, the ox, the swallow, and the bee. Certain plants—roses, heather, juniper, oak, it varied from village to village—were granted a special status. Some places in the landscape were seen as closer to the divine than others. Only with old age, on the part of some of the women virtuosi, were people able to move into and entirely inhabit this other world, brimming with excitement and meaning, in which every leaf that unfolded in the spring was a testimony to the power and love of God. Involved here is a sense of belonging, of communion with the landscape and habitat. That lexicon we discussed in the first chapter—the lexicon of persons' names and place names—had added to it the names, nicknames, and places of the divine. It was a part, perhaps the most essential part, of the village culture. To this world was added the wonderful mysteries of the Church, the divinely instituted infinity of devotions, and techniques to be investigated by the traditional means of observation, contemplation, and meditation. Thomas Merton writes, "The earliest fathers knew that all things, as such, are symbolic by their very being in nature, and all talk of something beyond themselves. Their meaning is not something we impose upon them, but a mystery we can discover in them, if we have but eyes to look." As an old man in Tudanca told me, "Everything in the world has its place, its assignment, although we do not know what they are." And my closest friend, old Antonia in San Sebastian, says, "Everything, even the tiniest thing, has its mystery."

There are degrees of belief in immanence. They form a continuum that runs from a world in which there is no coincidence to a world in which the laws of probability and randomness are fully operative. Most of the villagers fall, in respect to their beliefs, somewhere in the middle of the continuum. A few of the more traditional priests and many of the older villagers can decipher signs and portents everywhere. I am not at all sure that community centers and television sets adequately substitute for this degree of delight and wisdom. By their hardnosed rationalism the young priests live in a world where there is little or no participation of God in day-to-day affairs.

Another loss, for those in the society who cannot participate in peer groups, are the divine intermediaries. As we have seen time and again, lonely people find companionship in their personal patron, who provides them with comfort, reassurance, and love. Again this will hit older people, especially women, the hardest, but it will also affect the housewife isolated from her husband and the infirm isolated from the work cycle. The effect for those married may be somewhat mitigated by the fact that the young priests are also at work in changing the structure of marriage to make it egalitarian and affectionate.

Perhaps because it can be an instrument of unity, perhaps because they recognize that it is too sensitive an issue, the young priests have not, as of yet, moved on the village shrines and village patrons. The only case I know of is a young priest who had to fill in at the last moment as the preacher in his home village at the annual ceremony of Our Lady of Mt. Carmen. "I had no notes, but had some ideas I had been thinking about, and told the people that the saints and the advocations of the Virgin are found nowhere in the Bible, that salvation must come through Jesus Christ and that we must imitate him. I haven't been back to hear their reactions. I would not dare say that anywhere else, but I would in my own village." He went on, "In the Bible it says, 'we asked for bread and you did not give it to us.' We as priests are responsible for the parish; we cannot go on fooling the people, keeping the truth from them. Many times I have thought that if in Spain we said one rosary less and read the Bible more, we would be better off."

The seminaries are now almost wholly in the hands of the reform movement. If the pace of change is maintained as fast as at present, there may be a progressive removal from the liturgy of ceremonies to divine intermediaries. This would probably serve to alienate the priests even more from the villages they are attempting to change. They do regard the patronal images as a hindrance because of the way they limit the scope of the Christian society and the Christian family to the village or valley culture. They see the need for spreading the concept of familyhood to include all people, which might mean abandoning the idea of patronage on all but a universal scale. Since many of their parishoners will be moving to the city, since the village as a sacred landscape is already being abandoned and rejected for an urban or cosmopolitan culture by the villagers who still inhabit it, the priests are attempting to teach a more versatile, more universal religion that will not be limited in its applicability to a certain place.

What is to be feared is that the supports of the old society will be knocked away before the new society can be built. The priests by their sermons and by their spiritual direction in the confessional could destroy the Church's ideological support for the village philosophy long before they could have much concrete effects on social relations. In fact, in the case of the young priests I knew best, the impregnability of the village social structure and devotional system so discouraged them that they all went off to the cities, like Santander and Torrelavega, where a more fluid, dislocated social situation made their ideas much easier to put into effect, perhaps with a greater usefulness. But more young priests keep coming in, and one can only hope, for the sake of some of the villagers, that they act with patience and humaneness in making the great transformation.

APPENDIX I: DUTIES OF THE VARIOUS STATIONS

(A translation of p. 63-69 of Antonio Maria Claret,
Camino Recto y Seguro).

In order to examine the violations you may have committed against the obligations of your station, look up the one that applies to you in the Duties of the Various Stations which follow.

Duties of the Heads of Families

1. Maintain the family according to its proper station
2. Do not dissipate its possessions in gambling or frivolities
3. Pay the correct salary to servants, workers, etc.
4. Watch carefully over the behavior of your children and dependents
5. Be sure that they hear the word of God and receive the Holy Sacraments
6. Correct them with prudence
7. Punish them without wrath, etc.
8. Treat them with benevolence
9. Keep them busy
10. Help them in their illnesses
11. Edify them by your good example
12. Pray for them and supply them with good teachers, masters, etc.
13. Arrange for the correct separation between sons and daughters and persons of different sex
14. Admit no one who by their talk or in any other way would be a cause for scandal in the family

Duties of Children and Dependents

1. Regard and esteem your parents and masters as representatives of God
2. Love them
3. Respect them dutifully and speak well of them in their absence as well as in their presence
4. Obey them promptly
5. Serve them faithfully

6. Assist them in their necessities
7. Suffer their faults, always silently
8. Pray to God for them
9. Take care of household goods

Duties of the Husbands

1. Love your wife as Jesus Christ loves the Church
2. Do not scorn her, for she is your inseparable companion
3. Direct her as a subordinate
4. Take care of her, as you are her guardian
5. Maintain her with decency
6. Suffer her with patience
7. Help her with kindness
8. Correct her with good will
9. Do not mistreat her in words or in deeds
10. Do not do or say anything in front of your children, even when young, that might shock them

Duties of the Wives (see p. 31)

Duties of the Young

1. Be present at catechism
2. Respect the old
3. Avoid dangerous diversions
4. Flee from sloth and suspicious company
5. Do not go to bed late
6. Mortify the body
7. Avoid falling in love, profane songs, etc.
8. Do not take anything secretely, even from your own house
9. Pray to God and take the counsel of prudent men, to find the station in life that you should take

Duties of Damsels

1. Act on all occasions with consummate modesty
2. Speak very prudently
3. Do not desire to see or be seen
4. Do not dress with vanity
5. Avoid conversations alone with men

6. Abominate parties, dances, plays, etc.
7. Love pious exercises
8. Do not be idle even for an instant
9. Do some discreet mortification

Duties of Widows

1. Set a virtuous example for girls and married women
2. Friend of seclusion
3. Enemy of idleness
4. Lover of mortification
5. Given to prayer
6. Jealous of your good name

Duties of the Wealthy

1. Give thanks to God for your goods
2. Do not place your confidence in your possessions
3. Do not augment them by usury
4. Do not keep them unjustly
5. Do not use them to excite any passion
6. Be generous with the poor and with the Church
7. Ponder well that the rich are in great danger of condemnation because of the misuse of their riches

Duties of the Poor

1. Resign yourself to the will of God in your poverty
2. Do not take others' goods, even under pretext of poverty
3. Work hard to provide yourself with a modest well-being
4. Try to be rich with the things of eternity
5. Remember that Jesus Christ and Most Holy Mary were poor, too

Duties of Shopkeepers

1. Be content with a modest profit
2. Give the correct weight and measure to all
3. Do not falsify the merchandise
4. Do not monopolize one kind of good, causing the poverty of the people
5. Abstain from all fraud or cheating
6. Be generous with the poor

Duties of Artisans and Laborers

1. Offer to God often all your privations and fatigue
2. Work with total diligence and preciseness
3. Do not work on holy days; nor renege nor blaspheme
4. Do not keep the goods of others
5. Do not cause your employers to lose money
6. Do not lose time
7. Do not go back on your word
8. In work neither complain or have free conversations, etc.

After examining the conscience, and when you know the sins that you have committed, become genuinely sorry for them; if you don't, then you will be like that hunter who, having climbed hill and dale to flush out the game, neglects to shoot them when he finds them, and finds himself as tired as he is ridiculous. You will beseech God, then, by the intercession of the Most Holy Virgin, for the forgiveness of your sins, reciting to Him for this purpose six OUR FATHERS and seven AVE MARIAS, in memory of his sorrows; and doing acts of contrition and sorrow, you will say the following . . .

APPENDIX II: WITCHCRAFT AND OCCULT POWERS

In the early years of this century there were seers in these villages, women called *divinas*, who would help people out with their problems. They were able to call on special forces of perception to assist herdsmen in locating lost animals. Manuel Llano describes in detail the visit of a herdsman to a diviner in Cabuérniga, and I have had described to me diviners who helped people out in Tudanca, Santotís, and Sarceda. Apparently there is one who lives in Unquera now, and there may even be some divining still done in the villages that the people are unwilling to tell about. The diviners are older women. Some of them apparently work for a fee, others for nothing, except perhaps a gift. By using cards, or ribbons, and invoking the saints they are able to say what direction the animals have taken and whether they are dead or not. They are sometimes able to visualize the scene of where the animals are at the time. The diviners seem to be relatively benign supplements to the use of divine figures in time of crisis. I have the impression that they were used largely by men (who would be the herdsmen). But whether they were used instead of shrines, or after supplications to shrines had failed, I do not know. Now the custom is regarded as superstitious and is only discussed by the very old with any degree of seriousness.

The diviners, who are gifted local people, are not to be confused with traveling gypsies, who read palms for a fee. I talked to two people who had had their palms read by gypsies, and each had been amazed at the extent to which the gypsies' prophesies had come true.

There is some disagreement as to whether or not *divinas* are witches (*brujas*). The consensus of opinion is that they are not. Witches are considered to have the power of the devil at their disposal and are seen as malicious. As such even talking about them is sinful, and people are very reticent to do so. I was told only one story about witches:

> There were women in the village that people would say were witches, and there was something to say to protect yourself when they went by. There was a woman on her way to Santotís at night and a cat came and ran around her, back and forth in front of her, and she hit it with a stick, and the next day a woman was found who had broken her arm in bed at night. This happened before I was born, but I heard about it. (man, age 85, Tudanca)

This is almost precisely the same story that appears in Pereda's novel, *El Sabor de la Tierruca*. Pereda may have heard it in Tudanca, Tudanca may have gotten it from the book, or most likely, it is a common tale throughout the region. In the same book Pereda describes how the villagers (those of Polanco, near Torrelavega) assume that any isolated old woman, who lives alone and is slightly deformed, is a witch.[65]

Divinas also generally live alone. But the distinction between them and witches is made also in the Basque country, according to Caro Baroja.[66] It may be that some of the people who were considered witches were unsuccessful *divinas*. This could explain the confusion in the village between the two.

Attitudes toward women are implicit in the ascription of both benign and malign occult powers to them. Women are sources of uncertainty in the society. They have a tap on the unknown. Perhaps this is another reason for the use of Mary as the most important divine figure. Women who are married or living with others have their power somehow neutralized perhaps by the sexual act. Ancient etiologies of madness (Galen) still practiced in some shrines in Spain in the twentieth century (La Balma) ascribed hysteria to movement of the womb and prescribed manipulation of the clitoris in the shrine as a cure.[67] Women living alone are not under control. Their tap on the unknown is not closed off.

In a sense, by imputing impurity to women, the men obtain in the women, through the Church, a measure of self-regulation, of self-control. That this is necessary is demonstrated by popular beliefs that women are not only more likely to traffic with the devil (witchcraft) but also are most susceptible to possession by devils. The lists of miraculous cures of people possessed by demons given in the survey of shrines in Spain published in 1724 (2nd ed., 1740) shows three times as many women possessed by the devil as men. And it is to be noted that a distinction is made between possession and madness (*locura*). Men are more likely to be considered simply mad, women possessed.[68]

The only cases of possession in the valley that I was told about were told by an older woman in San Sebastian:

> Once in Lamason I remember hearing of a priest who during the fiesta told the story of a boy in a church in his home village who during the consecration of the host ran out and cried, "No quiero, no quiero, no quiero." (I don't want it.) The mother ran out to find what was wrong. The boy was possessed by the devil.

> And once in San Sebastian a family heard a great deal of ruckus in the stable under the house and found the animals acting strangely at night. They went to the priest who said they were possessed and then went and read the Gospels to the animals. But it was a secret kept within the family. They were supposed to tell no one else.

In both these cases the diagnosis of possession was given by the clergy. The cure of exorcism by reading the gospels is a common one. At the Dominican shrine of Las Caldas, little embroidered cloth amulets made by nuns and containing tiny snippets of paper from each of the four gospels are still sold. They are called *los Evangelios*. There are used to ward off evil spirits and also for cures. The bed in which I slept in San Sebastian had amulets of this kind hanging on the bedstead. I asked why they were there on two separate occasions. The first time I was told, "against witches and such like." And the second time I was told the whole story.

> One time I noticed a discolored spot on the wrist of one of my children, and I asked a *señora* who was in the house what the matter was, and she said it was probably some animal like a rat, and the way to protect against it was to put the *Evangelios* on the bed. I bought them from the nuns in Cabezon, and they kept the animals away.

There may be implicit in this story (considering the previous reference to witches) the idea that a witch in the form of a rat sucking her son's blood. Caro Baroja mentions rats as one of the species of animals (like cats in the earlier story) that witches commonly change into.

Another manifestation of the dark powers seems to be the evil eye. Certain persons are considered to have the ability to immobilize animals or other persons with their gaze. The verb for this operation is *entronizar*. One woman told of a case in which animals in a circus in Oviedo were bewitched by a woman in the audience and would not move until she left. And some people wondered whether the girls in the apparitions in San Sebastian, who sometimes fell into rigid, trancelike positions, were being held in place by someone's evil eye. On the whole, though, the evil eye is not a major preoccupation in these villages.

In fact to all of these phenomena the people in the valley have very mixed attitudes. They speak of them as though they had happened in the past, but do not happen now. At the end of a long talk about witchcraft with the lady in San Sebastian, she said, "We don't believe in witches. And if we do, we don't talk about them. God kicked the anti-Christ out of heaven, and with him his companions that we call devils. It doesn't seem as though we have cases of possession any more. But if we did we wouldn't hear about them." It was a very upsetting topic to bring up. She forgot about lunch. She kept saying that she wasn't going to talk about it, then thinking of something new to say. A dark corner.

In his *Paysans de Languedoc*, Emanual le Roy Ladurie has some remarkable pages in which he sketches (with particular reference to the Cathari) the idea that as people in Languedoc learned the public Catholic cult they also acquired (by innuendo, opposition, and secret whispering) a countercult of magic, occult forces, and vengence which they could call on in times of crisis or despair. That similar beliefs were held in these villages in the past is indisputable, but it is difficult to say whether whatever there was could be said in any sense to constitute a cult. The only things I have been able to discover, such as those given above, are fragmentary. It is impossible to say how well the theories of demonic or occult powers were elaborated. The present fragmentation is doubtlessly due both to the substitution of other sources of help in times of crisis and the anathema of the Church on the subject, which has rendered it almost entirely unspeakable.

BIBLIOGRAPHY

Adams, H. (1961). "Mont-Saint-Michel and Chartres." Mentor, New York.

Ames, M. M. (1966). Ritual Prestations and the Structure of the Sinhalese Pantheon, "Anthropological Studies in Theravada Buddhism." (M. Nash, ed.), Cultural Rep. Ser. No. 13, Southeast Asia Studies. Yale Univ. New Haven, 1966.

Anderson, T. Garelick, A. and Rubin, G. (1971). Workers Paid Off in Thing Called Love. Manuscript

Anonymous (1968). "Miguel Bravo." Editorial Bedía, Santander.

Artola, M. (1967). *La España del Antiguo Régimen.* Fasc. 3: Castilla La Vieja. *Acta Salmanticensis* Filosofia y Letras No. LV.

Attwater, D. (ed.) (1961). "A Catholic Dictionary," 3rd ed. Macmillan, New York; (1965). "The Penguin Dictionary of Saints." Penguin, London.

Barru, Rev. S. (1893). "Cabalistic Magic in Mediterranean Shrine Worship." Blockman, London.

Berger, P. L. (1969). "A Rumor of Angels; Modern Society and the Rediscovery of the Supernatural." Doubleday, Garden City, New York.

Berger, P. L. (1969). "The Sacred Canopy; Elements of a Sociological Theory of Religion." Doubleday, Garden City, New York.

Berger, P. and Luckmann, T. (1966). "The Social Construction of Reality." Anchor, New York.

Boutellier, M. (1966). "Médecine Populaire d'Hier et d'Aujourdi'hui." Maisonneuve et Larose, Paris.

Bremond, H. (1968). "Histoire Littéraire du Sentiment Religieux en France," Vol. 9. Armand Colin, Paris.

Buckner, N. A. (1930). "Le Sanctuaire et la Famine." La Trouille, Paris.

Caro Baroja, J. (1968). "Las Brujas y su Mundo." Alianza Editorial, Madrid.

Carrier, H. and Pin, E. (1967). "Essais de Sociologie Religieuse." SPES, Paris.

Chelhod, J. (1964). "Les Structures du Sacré chez les Arabes." G.-P. Maisonneuve et Larose, Paris.

Christian, W. A., Jr., Gale, S. and Wylie, J. (1970). An Introduction to the Ecology of Shrines in Spain, Working Paper #54. The Center for Research on Social Organization of the Univ. of Michigan. Ann Arbor, Michigan. Mimeographed.

Claret, A. M. (1946). "Camino Recto y Seguro para llegar al Cielo," 174th ed. Editorial Conculsa, Madrid.

Coe, S. D. (1963). The Catholic Saints of the Northwest Mediterranean. Unpublished Phd thesis, Dept. of Anthropology, Harvard Univ., June.

Cossio, J. M. de, and Maza Solano, T. (1920). "Cancionero Popular de la Provincia de Santander." Santander.

de la Hoz Teja, J. (1949). "Cantabria Por Maria." Centro de Estudios Montañeses, Santander.

de la Hoz Teja, J. (1951). "El Clero Montañés." Editorial Cantabrica, Santander.

Deutsch, K. (1966). "Nationalism and Social Communication." MIT Press, Boston, Massachusetts.

Díez Llama, S. (1971). "La Situación Socio-Religiosa de Santander y el Obispo Sanchez de Castro (1884-1920)." Institución Cultural de Cantabria, Santander.

Diputación Provincial de Santander. (1944-1945). "Investigación de las Riquezas rústicas y pecuarias de la provincia de Santander." Santander.

Dominguez Ortiz, A. (1955). "La Sociedad Española en el Siglo XVIII." CSIC, Madrid.

Douglas, M. (1966). "Purity and Danger; An Analysis of Concepts of Pollution and Taboo." Praeger, New York.

Du Manoir, H. (1956-1964). "Maria," 7 vols. Beauchesne, Paris.

Duocastella, R., et al. (1967). "Análisis Sociológico del Catolicismo Español." ASPA, Madrid.

Durkheim, E. (1965). "The Elementary Forms of the Religious Life." Free Press, New York.

Edelstein, E. J., and Edelstein, L. (1945). "Asclepius; A Collection and Interpretation of the Testimonies," 2 Vols. Johns Hopkins Press, Baltimore, Maryland.

Freeman, S. T. (1968). Religious Aspects of the Social Organization of a Castillian Village, *Amer. Anthropologist* **70**, No. 1, 34-49.

Geertz, Clifford. (1968). Religion as a Cultural System, "Anthropological Approaches to the Study of Religion" (M. Banton, ed.), Travistock, London.

González, J. M. "Toponomia de una Parroquia Asturiana." CSIC, Oviedo.

González Echegaray, J. (1969). "Orígines del Cristianismo en Cantabria." Institución Cultural de Cantabria, Santander.

Gougaud, D. L. (1922), Anciennes traditions ascétiques, *Rev. Ascétique Mystique* **3**, 56-59; (1923). 4 140-156.

Graef, H. (1963, 1965). "Mary; A History of Doctrine and Devotion," 2 vol. Sheed and Ward, New York.

Harding, S. (1970). Ensayo Crítico sobre la vida ganadera. Manuscript, Zaragoza.

Hinton, W. (1966). "Fanshen." Vintage, New York.

Instituto Nacional de Estadística (1965). Reseña estadística de la Provincia de Santander. Madrid.

James, W. (1958). "The Varieties of Religious Experience." Mentor, New York.

Jaslow, J. (1970). The Poor in Church. Manuscript.

Kendrick, T. D. (1960). "St. James in Spain." Metheun, London.

Kenny, M. (1966). "A Spanish Tapestry." Harpers, New York.

Lacotte, R. (1953). "Recherches sur les Cultes Populaires dans l'actuel diocese de Meaux" (Seine et Marne). Memoires de la Fédération Folklorique de l'Île de France, Paris.

Leach, E. R. (ed.) (1968). "Dialectic in Practical Religion," Cambridge Papers in Social Anthropology No. 5 Cambridge Univ. Press, London and New York.

Léal, J. (1961). "Año Cristiano," Madrid.

Leproux, M. (1957). "Dévotions et Saints Guerisseurs." Presses Univ. de la France, Paris.

Le Roy Ladurie, E. (1966). "Les Paysans de Languedoc," 2 vols. S.E.V.P.E.N., Paris.

Levi-Strauss, C. "The Savage Mind." Univ. of Chicago Press, Chicago, Illinois.

Lewis, R. W. B. (1955). "The American Adam; Innocence, Tragedy, and Tradition in the Nineteenth Century." Univ. of Chicago Press, Chicago, Illinois.

Linz, J. and de Miguel, A. (1966). Within-Nation Differences and Comparisons: The Eight Spains, "Comparing Nations" (R. Merritt and S. Rokkan, eds.). Yale Univ. Press, New Haven, Connecticut.

Lison-Tolosano, C. (1966). "Belmonte de los Caballeros." Oxford Univ. Press, London and New York.

Lizuriaga, L. (1926). "Analfabetismo en España." Cosano, Madrid.

Llanos, M. (1968). "Obras Completas," 2 vols. Santander.

Madden, M. R. (1930). "Political Theory and Law in Medieval Spain." Fordham Univ. Press, New York.

Maza Solano, T. (ed.) (1957). "Aportación al Estudio de la Historia Economica de la Montaña." Santander: Centro de Estudios Montañeses, 1957.

Maza Solano, T. (ed.) (1965). "Relaciones Histórico-Geograficas y Económicas del Partido de Laredo en el siglo XVIII." Centro de Estudios Montañeses, Santander.

Maza Solano, T. (ed.) (1953-61). "Nobleza, Hidalguia, Profesiones y Oficios en la Montaña, según los Padrones del Catastro del Marqués de la Enseñada." Centro de Estudios Montañeses, Santander.

Menendez Pidal y Alvarez, L. (1958). "La Cueva de Covadonga; Santuario de Nuestra Señora la Virgen Maria." Instituto de Estudios Asturianos, Oviedo.

Molino Navarrete, J. del (1681). "Constituciones Añadidas a las Synodales de Palencia." por Antonio González de Reyes, Madrid.

Morin, E. (1970). "The Red and the White." Pantheon, New York.

Mumford, L. (1961). "The City in History." Harcourt, Brace and World, New York.

Nash, M. (ed.) (1966). "Anthropological Studies in Theravada Buddhism," Cultural Rep. Ser. No. 13, Southeast Asia Studies. Yale Univ. Press, New Haven, Connecticut.

Neal, Sister Maria Augusta. (1965). "Values and Interests in Social Change." Prentice-Hall, Englewood Cliffs, New Jersey.

Nistal, A. (1965). "La Virgen Bien Aparecida; Patrona de la Montaña." Salamanca.

Obeyesekere, G. The Buddhist Pantheon in Ceylon and its Extensions, "Anthropological Studies in Theravada Buddhism." (M. Nash, ed.). Yale University Press, New Haven, Connesticut.

Obeyesekere, G. (1968). Theodicy, Sin and Salvation in a Sociology of Buddhism, "Dialectic in Practical Religion" (E. Leach, ed.), Cambridge Univ. Press, London and New York.

Obregón, E. (1971). "Santander 1937-71; Planteamientos para la Historia de una Diócesis." Editorial Bedía, Santander.

Oficina General de Estadística y Sociología Religiosa, and Departmento de Investigación Socioreligiosa. (1968). Anteproyecto de Temas de Estudio. Madrid. mimeographed, n.d.

Otto, R. (1958). "The Idea of the Holy." Oxford Univ. Press, London and New York.

Pereda, J. M. de (1960). "Peñas Arriba." Expasa-Calpe Argentina, Buenos Aires.

Pereda, J. M. de (1960). "El Sabor de la Tierruca." Espasa-Calpe Argentina, Buenos Aires.

Pérez, N. (1940-1949). "Historia Mariana de España." 5 vols. Gráficas J. Concejo and Impresos Gerper, Valladolid.

Pin, E. (1967). Visions Religiéuses du Monde en Amerique Latine, "Essais de Sociologie Religéuse" (H. Carrier and E. Pin, eds.), Paris.

Pitt-Rivers, J. (1961). "The People of the Sierra." Univ. of Chicago Press, Chicago, Illinois.

Prats y Beltran, A. (1929). "Tres Días con los Endemoniados." Editorial Cenit, Madrid.

Rappaport, R. A. (1967). "Pigs for the Ancestors; Ritual in the Ecology of a New Guinea People." Yale Univ. Press, New Haven, Connecticut.

Revenga Carbonell, A. (1960). "Comarcas Geográficas de España." Instituto Geográfico y Catastral, Madrid.

Richardson, M., Bode, B., and Pardo, M. E. (1969). The Image of Christ in Spanish America as a Model for Suffering: An Exploratory Note, mimeographed, n.d. Louisiana State Univ.

Sahlins, M. D. (1968). "Tribesmen." Prentice-Hall, Englewood Cliffs, New Jersey.

Sainz de los Torres, M. (1906). "Breve Reseña de los Santuarios Marianos en la Provincia de Santander." Sucesores Rivadeneyra, Madrid.

Sanchez Perez, J. A. (1943). "El Culto Mariano en España." CSIC, Madrid.

Scully, V. (1969). "The Earth, The Temple, and The Gods," revised ed. Praeger, New York.

Slicher Van Bath, B. H. (1963). "The Agrarian History of Western Europe." Arnold, London.

Smart, N. (1958). "Reasons and Faiths; An Investigation of Religious Discourse, Christian and Non-Christian." Routledge and Kegan Paul, London.

Staehlin, C. M. (1954). "Apariciones." Razón y Fe, Madrid.

Stirling, P. (1963). The domestic cycle and the distribution of power in Turkish villages, "Mediterranean Countrymen; Essays in the Social Anthropology of the Mediterranean." (J. Pitt-Rivers, eds.). Mouton, Paris, the Hague.

Swanson, G. E. (1967). "Religion and Regime: A Sociological Account of the Reformation." Univ. of Michigan Press, Ann Arbor, Michigan.

Swanson, G. E. (1968). To Live in Concord with Society, "Cooley and Sociological Analysis" (A. J. Reiss, Jr., ed.). Univ. of Michigan Press, Ann Arbor, Michigan.

Tentler, T. "Forgiveness and Consolation in the Religious Thought of the late Middle Ages." (forthcoming)

Titiev, M. (1960). A Fresh Approach to the Problem of Magic and Religion, *Southwestern J. Anthropol.* **XVI:3** 292-298.

Valuy, B., S. J. (1924). "Directorio del Sacerdote en su vida privada y pública." Apostolado de la Prensa, Madrid.

Vazquez, J. M., (1967). "Realidades Socio-religiosas de España." Editora Nacional, Madrid.

Vieth, I. (1965). "Hysteria; The history of a disease." Univ. of Chicago Press, Chicago, Illinois.

Vilafañe, J. de. (1740). "Compendio Histórico en que se da noticia de las milagrosas, y devotas imagenes. . . .," 2nd ed. Salamanca.

Viñayo Gonzalez, A. (1964). "La Devoción Mariana en Asturias," Arch. Leoneses **34**.

Wolf, E. R. (1969). Society and Symbols in Latin Europe and in the Islamic Near East: Some Comparisons. *Anthropolog. Quart.* **42**, No. 3. 287-301.

Wylie, L. (ed.) (1968). "Chanzeaux, A Village in Anjou." Harvard Univ. Press, Cambridge, Massachusetts.

FOOTNOTES

[1] Census of 1870 and L. Lizuriaga, "Analfabetismo en España." Madrid, 1926.

[2] M. Llano, "Obras Completas," Vol. 1, Santander, 1968. p. 362.

[3] See the Spring, 1963 issue of Anthropological Quarterly devoted to the cultural ecology of Western Europe.

[4] P. Branche, Ces forêts qui mangent les champs. *Le Monde* p. 9. Paris (August 12, 1970).

[5] M. Llano, "Obras Completas," Vol. 1, p. 191. Santander, 1968.

[6] Most of my ideas on identity ultimately derive from K. Deutsch, "Nationalism and Social Communication." Boston, Massachusetts, 1966.

[7] The major sources for the history of the region are the publications of the Centro de Estudios Montañeses, under the direction of Tomas Maza Solano.

[8] J. M. Gonzalez, "Toponomia de una Parroquia Asturiana." Oviedo, 1959.

[9] From the works of M. Llano it seems that in Carmona at the beginning of this century the *mocedad* was a male group that organized the town fiestas and made rounds of the village singing songs under the girls' balconies. It was apparently more formally organized then, with a leader called "el caporal" who informed the chavales when they had been admitted.

[10] J. M. de Pereda, "El Sabor de la Tierruca," p. 119. Buenos Aires, 1960.

[11] A. M. Claret, "Camino Recto y Seguro para llegar al Cielo," pp. 63-69. Madrid, 1946.

[12] J. M. Pereda, "El Sabor de la Tierruca," p. 97. Buenos Aires, 1960.

[13] E. Morin, "The Red and the White." New York, 1970; W. Hinton, "Fanshen." New York, 1966.

[14] A. Dominguez Ortiz, "La Sociedad Española en el Siglo XVIII," p. 103. Madrid, 1955.

[15] D. Ringrose, Perspectives on the Economics of 18th century Spain. Paper delivered at the Society for Spanish and Portuguese Historical Studies, April 17, 1971, State Univ. of New York at Stonybrook.

[16] P. Baroja, "The Restlessness of Shanti Andia," p. 114. New York, 1962.

[17] Evidence that in depressions in western Europe peasants have tended to stay at home will be found in B. Thomas, "Migration and Economic Growth." Cambridge, England, 1954; S. Kuznets and D. S. Thomas, "Population Redistribution and Economic Growth." Philadelphia, Pennsylvania, 1957; L. Wylie, ed., "Chanzeaux, A Village in Anjou." Cambridge, 1968.

[18] The table is of marriages in the parish marriage registries. Since bethrothed couples from different villages marry normally in the village of the bride, this table is incomplete for those men from the villages in question who married outside their home village. Part of the rise in marriages with men from neighboring valleys at the turn of the century is because men from these valleys worked on the highway that was built in the valley at that time. Similarly, from 1940 to 1945 many girls in the valley married outsiders who were temporarily in the valley to build the Saltos del Nansa dam. The tables on the whole probably also reflect the pattern of girls working away from the valley, a custom which gained momentum at the turn of the century and became common after the war.

[19] A. Revenga Carbonell, "Comarcas Geográficas de España." Madrid, 1960.

²⁰ P. de Gorosabel, "Noticia de las cosas memorables de Guipúzcoa," 2nd ed., Vol. 2, pp. 430-431. Bilbao, 1967.

²¹ Information on the shrine of La Velilla is available in leaflets distributed at the shrine; J. de Vilafañe, "Compendio Histórico . . . ," pp. 592-595, Salamanca, 1740; J. A. Sanchez Perez, "El culto mariano en España." Madrid, 1943.

²² The chief source on Covadonga is L. Menendez Pidal y Alvarez, "La Cueva de Covadonga; Santuario de Nuestra Señora La Virgen Maria." Oviedo 1958. Ant. Viñayo failed to find any mention of the shrine in his survey of documents relating to Marian devotion in the province in the first five centuries of the reconquest (La Devocion Mariana en Asturias, "Archivos Leonesas XXXIV." Leon, 1964).

²³ J. de Vilafañe, "Compendio Histórico . . . ," p. 138. Salamanca, 1740.

²⁴ For El Brezo see "Historia del Santuario de Nuestra Señora del Brezo," Burgos, 1939; and Sanchez Perez (footnote 21).

²⁵ M. Sainz de los Torres, "Breve Reseña de los Santuarios Marianos en la Provincia de Santander." Madrid, 1906.

²⁶ A. Nistal, "La Virgen Bien Aparecida; Patrona de la Montaña," pp. 54-55. Salamanca, 1965.

²⁷ The rise of Bielba also coincides with the decline of St. Toribio, where the feast of the Holy Cross is celebrated on the same day. Bielba doubtless preempted visits to more distant Potes on the part of the Obeso villagers.

²⁸ "Reglamiento y Devocionario de la Hermandad del Stmo. Cristo de los Remedios de Bielva," pp. 5-6. Santander, 1918.

²⁹ "Nuestra Señora de la Luz de Liébana venerada en las Montañas de Peña Sagra; Historia; Novena; y Auto." Santander, 1955.

³⁰ S. Dobzhansky Coe, The Catholic Saints of the Northwestern Mediterranean, pp. 102-103. Unpublished Ph.D. thesis, Dept. of Anthropology, Harvard Univ. (June 1963).

³¹ G. Obeyesekere, The Buddhist Pantheon in Ceylon and its Extensions, in "Anthropological Studies in Theravada Buddhism" (M. Nash, ed.), p. 25. New Haven, Connecticut, 1966.

³² Libro de Fabrica, Ermita de Nuestra Señora del Llano, Obeso.

³³ Many other images in Spain go to parish churches from their chapels for stays of varying lengths during the year. In some cases, like La Luz on Peña Sagra, the reason seems to be to permit devotion while the chapel is snowbound (cf. Peña de Francia and Moncayo). In other cases, like the Virgin of Tiscar, near Quesada, Jaen, the image visits the parish church in order to permit, as at Tudanca, a more intense, personal devotion.

³⁴ L. Mumford, "The City in History," p. 280, New York, 1961.

³⁵ A. Dominquez Ortiz, "La Sociedad Española en el Siglo XVIII," p. 146. Madrid, 1955.

³⁶ A. Dominquez Ortiz, "La Sociedad Española el Siglo XVIII," p. 162. Madrid, 1955.

³⁷ D. Attwater, ed., "A Catholic Dictionary," p. 442. New York, 1961.

³⁸ T. D. Kendrick, "St. James in Spain." London, 1960.

³⁹ J. de la Hoz Teja, "Cantabria Por Maria," pp. 205-215. Santander, 1949.

⁴⁰ It seems that for some Catholics the apparitions at San Sebastian took the place of this third secret message.

⁴¹ The appearance of Our Lady of Mt, Carmel in San Sebastian was preceded by the appearance of Saint Michael the Archangel. The pair of them are copatrons of Cosío, the nearest village to San Sebastian down the mountain.

⁴² The maize was brought throughout the year to the house of the majordomo for each saint, then on a Sunday in May, it was auctioned off at the church after mass. The maize given was on the basis of promises made and favors received from the saint in

question. The system served to both raise money for the church and also to redistribute maize from the families that had enough to offer it to those who needed it and could bid on it at the auction. Similar systems exist among certain South American Indian tribes. May is the time when a lack of maize would begin to be felt by those running short. Probably one of the reasons that the custom has virtually died out is that maize is no longer the staple food of the culture, and its redistribution therefore no longer is important. Note that the redistribution took place under the beneficent auspices of the saints. This method of collection and redistribution mirrors the methods of redistribution of the family, which points to the role of the saints as parents of the community.

[43] My chief sources on the saints in Spain were: J. Léal, "Ano Cristiaño." Madrid, 1961; also D. Attwater, "The Penguin Dictionary of Saints." London, 1965.

[44] It is difficult to accurately deduce devotions from the use of baptismal names, but they do shed some light on the popularity of certain saints up to the nineteenth century, and on the popularity of Mary up to the present. The problems are these: In addition to special devotion to a given saint, there are other reasons for choosing a Christian name. One might give the child the surname of the father, of some other relative, or of the godparents, especially because the godmother often chooses the name. At least in recent years, this does not seem to have been a common custom, however. Furthermore, even if it were, it would not distort the statistics drastically since children are often given two, even three names. A bigger problem, in very recent years, may be the giving of names of television personalities or pop singers. Another problem, which seems to have been particularly strong from 1806 until 1900, is the custom of strictly naming the child after the saint on whose day he was born and giving him or her no other additional names. This practice virtually disqualifies those years from consideration. I do not know whether it was a policy of the Church, or merely a fad but from the names of characters in novels of that period I know that the custom was general at least across the north of Spain.

Therefore, we summarize that some information on popular devotions can be inferred from those baptismal names used before 1806; those used in the nineteenth century more commonly than chance would indicate; and those used during the twentieth century.

There were several ways to tabulate the findings. First of all, I counted only the most commonly used names (15 male and 15 female) by 10-year periods. I had taken down all the names given in San Sebastian from 1638 to the present, in Tudanca from 1858 to the present, and in Santotís and La Lastra from 1926 to the present. Then I counted all the names given, all the names given to males, and all the names given to females. In order to have a long series of percentages I based the frequencies of any given name upon the total number of names given, not the total number of persons. Basing them on the number of persons would have thrown the series askew when the custom changed as to the number of given names, for any name would be more likely to occur on a given person when the custom was to give more than one name. As an exception to this method of calculation, I compiled separate percentages on the frequency of Mary of Marian advocation (e.g., Dolores, or Carmen). Whereas any saint's name might or might not be the primary name, the Mary name almost always came before any other incidental names. I felt safe in assuming, therefore, that if there had only been one name, it generally would have been the Marian name. Hence while the calculation of saints' names are for the percentage of total names with a given name, those for Mary are the percentage of women with Marian name. Incidentally, it is fascinating that a new cycle of Marian apparitions in Spain should occur as the curve of Marian devotion, as represented by these baptismal names, is undergoing its maximum rate of ascent.

[45] The rank orders are better for comparisons of relative popularity over time because the percentages vary according to the number of names given in the epoch per child (as well as the aforementioned criteria for giving names in the first place). Until about 1710 only a

single name was given at baptism. Afterward, until about 1750, two were often given. And from 1750 to 1806 it was not unusual to have three given names. From 1806 until 1851 the number of names was usually only one, again, but since 1851 persons have been given either one or two names.

[46] D. Attwater, ed., "A Catholic Dictionary," p. 413. New York, 1961.

[47] Claret, footnote 11, p. 413.

[48] Claret, footnote 11, p. 420.

[49] "Libro de la Cofradia de Las Benditas Animas." Sarceda.

[50] See R. M. Marie de la Visitation, Marie et le Purgatoire, in "Maria; Etudes sur la Sainte Vièrge" (Hubert Du Manoir, ed.), Vol. V., pp. 887-923.

[51] Here is a list of the brotherhoods of the rosary, probably incomplete, with the dates of their foundings and refoundings: Obeso, 1689; Tudanca, before 1701, 1791; Cabrojo (=Puentenansa), 1876; Sarceda, 1892; Aniezo (Liébana), 1762, 1893.

[52] According to Dominguez Ortiz (footnote 14, p. 154) the missions were begun in the second half of the seventeenth century.

[53] See p. 121.

[54] B. Valuy, "Directorio del Sacerdote en su Vida privada y publica," pp. 74-75. Madrid, 1924.

[55] D. Attwater, ed., "A Catholic Dictionary," p. 98. New York, 1961.

[56] Gananath Obeyesekere, Theodicy, Sin and Salvation in a Sociology of Buddhism, in "Dialectic in Practical Religion" (Edmund Leach, ed.), p. 30. Cambridge, 1968.

[57] On a more universal sociopsychological level, another answer may lie in the argument that the anthropologist Mary Douglas might make, that with the continued existence of these little societies becoming increasingly problematic—their boundaries and identities threatened—there occurs an obsessive preoccupation with all boundaries, those of the body as well as those of the society; that sex, menstruation, and childbirth are seen as polluting violations of these boundaries; that the emigrants (see below) with their ambiguous allegiance and their dubious adherence when absent to the societal ethics are threatening the external boundaries of the society, for which they too feel a sense of impurity; that there is *no* sense, on the other hand, in which the profanity of the life of self-interested activity threatens the intactness or the consistency of the social system, and therefore no way it would generate a sense of impurity.

[58] I am grateful to Juan Linz for the electoral maps by *partido judicial* that permit this statement.

[59] J. del Molino Navarrete, Obispo de Palencia, *"Constituciones Añadidas a las Synodales de Palencia,"* p. 50. Madrid, 1681.

[60] J. del Molino Navarrete, footnote 59, p. 62.

[61] Marshall Sahlins has related different degrees of reciprocity (he distinguishes three: generalized, balanced, and negative) with different degrees of distance from the individuals for tribal situations. It is remarkable the degree to which his discussion applies here, how much the outlook of these villages resembles that of tribes. I had not read his work when I wrote the pages that follow. (See "Tribesman") Prentice-Hall, Englewood Cliffs, New Jersey, 1968.

[62] For an excellent study of the parallels between human-human and human-divine exchanges of goods see M. M. Ames, Ritual Prestations and the Structure of the Sinhalese Pantheon, *in* "Anthropological Studies in Theravada Buddhism" (M. Nash, ed.), pp. 27-50. New Haven, Connecticut, 1966.

[63] Claret, footnote 11, pp. 63-69.

[64] R. W. B. Lewis, *"The American Adam,"* p. 3. Chicago, 1955.

[65] Pereda, footnote 10, pp. 59-60.

⁶⁶ J. Caro Baroja, "Las Brujas y su mundo," pp. 280-298. Madrid, 1968.

⁶⁷ See Ilza Vleth. Hysteria, "The History of a Disease," pp. 38-39. Chicago, 1965, and A. Prats y Beltran, *"Tres Dias con los Endemoniados,"* Madrid, 1929.

⁶⁸ Vilafañe, footnote 21, *passim.*

SUBJECT INDEX

PERSON AND GOD IN A
SPANISH VALLEY

Focusing on long neglected doctrines, practices, and wisdom below the level of orthodox
Catholic dogma, this outstanding example of
contemporary social science will add a new
dimension to the anthropological and sociological study of religion. In addition, it will be of
great value to Catholic clergy and laity who are
concerned with the impact of the reforms of
Vatican II on the values of the old Church.

The book is a study of the religious life of the
people of the Nansa valley of northern Spain —
site of apparitions of the Virgin and St. Michael
in the early 1960's — and is based on a year
spent sharing their life. The author believes that,
by understanding the roles and relationships
that characterize a social group, one can begin
to understand the group's religious experience.
Accordingly, the book first describes the people
themselves, their society, and the annual round
of activities in this herding culture in the
Cantabrian mountains.

The second section describes the divine figures
that make up the pantheon of these people and
the three forms of religion which they practice.
The first and oldest (perhaps even antedating
Christianity) is manifest in many shrines which
exert their influence over specific areas (nation,
province, vale, village) that correspond to the
levels on which the people form a community
or have a sense of identity. The shrines are used
to deal with the concrete problems of life — i.e.
as loci for soliciting divine energy for human
purposes and eliciting human energy for divine
purposes. Another, more personal, form of
religion — probably introduced after the Council
of Trent — is characterized by a sense of sin, a
fear of Purgatory, and salvation as a goal. It is
marked by generalized devotions, such as the
Sacred Heart and the rosary, and is concerned
with transforming people from one spiritual
condition to another. Finally, a third form has
been introduced by young priests following
Vatican II, a form in which the people are
encouraged to find God in one another and to
stop relying on divine intermediaries.

Continued on back flap.